Cap Barriers: Halt Slope & Landfill Issues

Michael

LIST OF CONTENT

CHAPTER 1

INTRODUCTION

1.1 Background

In the geotechnical engineering field, capillary barrier systems are used mainly for two reasons: to avoid slope instabilities and to reduce infiltration of rainwater into water-sensitive areas such as landfill sites. Without a capillary barrier system, the infiltration of water into unsaturated soil deposits can reduce the strength of soil and lead to slope failure. Similarly, a reduced amount of infiltrated water minimizes the transport of contaminants out of landfills. An example of capillary barrier systems built on a slope is shown in Figure 1.1.

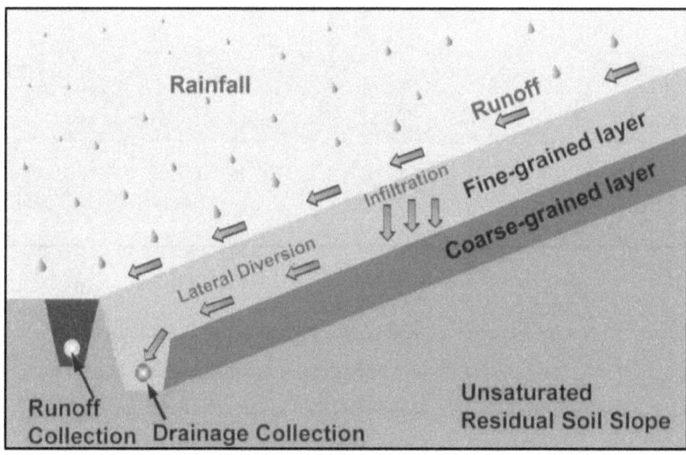

Figure 1.1 A sketch of a capillary barrier that can be built on a slope (from Rahardjo et al. 2012)

As shown in the above figure, the simplest capillary barrier system consists of a fine-grained soil layer placed over a coarse-grained soil layer. The fine-grained soil layer prevents or reduces the percolation of water into the coarse grained soil layer. Consequently, the unsaturated soil below the coarse-grained soil layer is protected from the rainfall. Capillary barrier systems make use of the principles of unsaturated soil mechanics to reduce rainwater infiltration and to prevent the water from reaching the natural soil deposits on a slope or the industrial waste in landfills. The overall goal of such systems is to minimize the generation of unwanted leachate.

One of the basic requirements for a capillary barrier system to function satisfactorily is a sufficiently large difference between the hydraulic properties of fine-grained and coarse grained soil layers. The relevant hydraulic properties for the design of a capillary barrier system are the soil water characteristics curves (SWCCs) and hydraulic conductivity functions of both fine and coarse soils as shown in Figure 1.2.

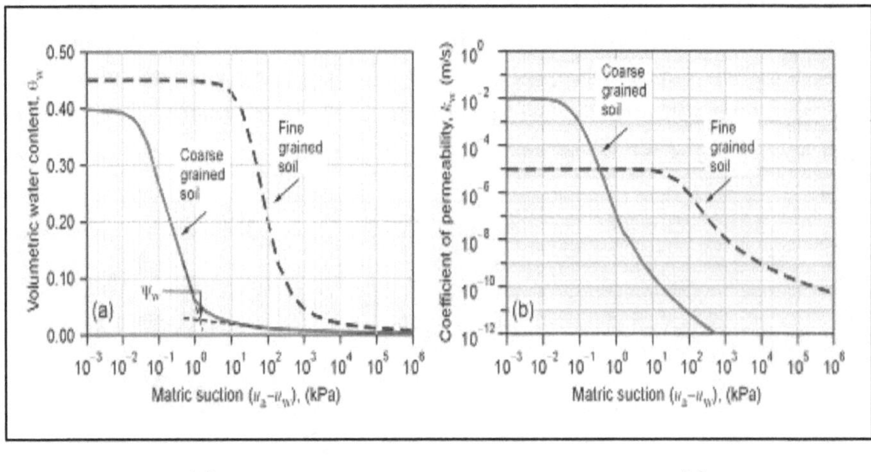

(a)   **(b)**

Figure 1.2 The SWCCs for fine grained and coarse grained soils: (a) volumetric water content vs. matric suction (ua-uw), and (b) coefficient of hydraulic conductivity vs. matric suction (ua-uw). (from Rahardjo et al. 2009). Note: the symbol ψ_w in Figure 1.2a indicates the water entry value WEV if the SWCCs were obtained from wetting type of experiments

The difference in hydraulic properties forms the hydraulic impedance, which limits the downward water movement. In other words, infiltrating water remains contained in the fine-grained soil layer and it does not move into the underlying coarse-grained soil layer until certain conditions are no longer satisfied. Fredlund and Rahardjo (2009) provided two design criteria for the capillary barriers to function properly:

"The first design criterion states that the matric suction at the interface between the fine and coarse soil layers must always remain higher than the intersection of permeability functions for the two materials. Failure to meet this criterion results in unwanted infiltration. The second design criterion states that the matric suction at the interface between the fine and coarse soil layers must be kept greater than the residual suction of the coarse layer." (Fredlund and Rahardjo 2009).

The step-by-step processes leading to percolation of water into coarse-grained soil layer can be described as follows. When a capillary barrier system is constructed, both soil layers are unsaturated and their water contents are low. In the case of prolonged precipitation and increasing infiltration, water begins to accumulate first within the fine-grained soil layer. Its volumetric water content and hydraulic conductivity gradually increase. With time, the infiltration would proceed through the fine-grained soil layer and reach the interface between the fine-grained and coarse-grained soil layers. As the matric suction at the interface approaches the water entry value (WEV) of the coarse-grained soil, the coarse-grained soil layer will begin to saturate rapidly and its hydraulic conductivity will increase. Once the hydraulic conductivity of the coarse grained soil layer becomes equal to or larger than that of the fine grained soil layer, the effectiveness of the capillary barrier will rapidly decrease. At this stage, percolation will occur (Wilson et al. 1995; Bussiere et al. 2003; Rahardjo et al. 2009).

Maintaining the moisture content in the coarse-grained soil layer at a low level at all times is an essential requirement for the capillary barrier to satisfy the two design criteria mentioned above. Stormont et al. (1998) circulated continuous atmospheric air flow through the coarse soil layer to maintain the moisture content at low levels. As a result, the performance of the capillary barrier remained satisfactory. Similarly, Baehr et al. (1989) described a technique for the removal of volatile contaminants from soil deposits. Air was forced to flow in the soil and the soil vapor was extracted during this operation. Pumps were used to force the air at one end and air and vapor were extracted using withdrawal wells at the other end of a contaminated unsaturated soil layer.

1.2 Research objectives

The main objectives of this investigation are:

1. To develop a method using heated air flow to prevent infiltration of water into the coarse-grained soil layer of capillary barriers.
2. To determine the impact of heated air flow on the water removal rate and recovery of capillary barrier systems which may have undergone a percolation breach event.
3. To develop a numerical model to simulate the experimental study and to extend the numerical work to case studies to demonstrate the benefits of heated air flow in the coarse-grained soil layer of capillary barrier systems.

1.3 Scope of the investigation

The objectives of this research are achieved by laboratory experiments and numerical analysis. Several model scale tests are conducted in the laboratory to investigate the effect of heated air flow on the moisture content and suction values of unsaturated soil mass contained in a sand box. The experimental setup and test results are given in Chapter 3. The finite element simulations of the experimental work are provided in Chapter 4. In Chapter 5, a relatively large scale laboratory test on a capillary barrier constructed on a sloping ground (Tami et al. 2004) is analyzed first using the finite element analysis. Subsequently, the analysis is extended to show the benefits of using heated air flow through the coarse grained soil layer. An actual case study on a capillary barrier system is discussed in Chapter 6. The potential benefits of using heated air flow are demonstrated using numerical analysis. The conclusions of this research are summarized in Chapter 7.

CHAPTER 2

LITERATURE REVIEW

2.1 Introduction

This section presents a literature review pertinent to dry capillary barrier systems. This topic is closely related to the research area of the present study. The concept of the dry capillary barrier systems, various design methodologies, and the methods for humidity control are discussed. In addition, a number of drainage techniques available in practice for capillary barrier systems are reviewed. Further, this chapter highlights the need for a new method to increase the efficiency of capillary barrier systems by either eliminating the percolation or delaying the percolation of water into waste material. The topic of transient moisture flow and heat transfer in unsaturated soils is also covered in some detail. At the end of the chapter, the thermodynamic behaviour of capillary barrier systems is discussed from a theoretical perspective (i.e., coupled moisture flow and heat transfer in unsaturated soils).

A capillary barrier system is an engineering solution that has been used for various civil engineering projects. Capillary barriers are frequently used as a soil cover for landfills and mining waste to reduce infiltration of water into waste materials. They are also used as a method of slope stabilization by preventing infiltration of rainwater into the native soil mass. Excessive amount of water may induce landslides. The soil layers in a capillary barrier consist of fine and coarse materials, which are typically unsaturated. The preferred configuration is usually a fine soil layer overlying a coarse soil layer. The system is also applicable to other geotechnical and environmental engineering structures, such as road embankments (Bussiere et al. 2000; Khire et al. 2000; Tidwell et al. 2003; Yang et al. 2004; Krisdani et al. 2005; Rahardjo et al. 2009). In this chapter, some background information is provided to better illustrate the physics involved in the processes related to capillary barrier systems.

The term "capillary" in the system's name refers to the capillary action, in other words, the tension forces between the two opposing layers (Krisdani et al. 2005). The capillary action will lead to water being held by matric forces and capillary menisci developed above the interface between the fine-grained soil layer and the coarse-grained soil layer (Mancarella and Simeone,

2012). Surface tension and attractive forces between the soil ions and the water molecules will develop as shown in Figure 2.1.

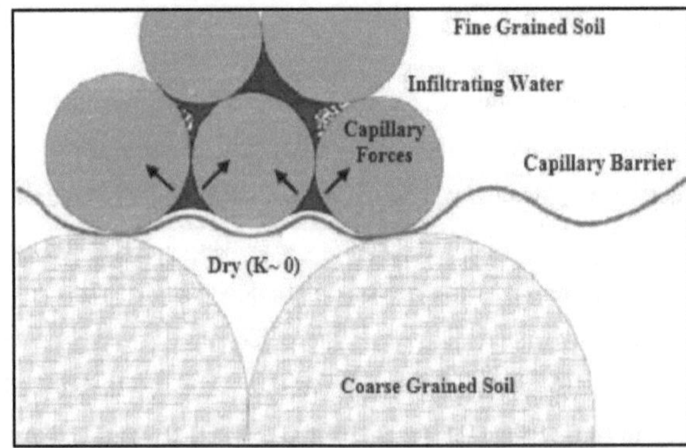

Figure 2.1 Phenomena of the capillary action (Mancarella and Simeone 2012)

Many researchers (e.g., Morris and Stormont 1999; Bussiere et al., 2003) have indicated that the capillary barrier system is a simple, durable, and a low construction cost solution to a number of geotechnical problems.

2.2 Principles of operation

A capillary barrier takes advantage of the higher capillary rise in the fine-grained soil compared to that of the coarse-grained soil to prevent water migration downward through the interface between the fine-grained soil layer and the coarse-grained soil layer. The difference in the unsaturated coefficients of hydraulic conductivity of the fine and coarse grained soil layers in the barrier is what creates the "barrier" effect. Another contributor to this effect is the difference in the matric suction between the two layers. The matric suction value for the fine soil layer is higher than that for the coarse grained soil layer because of the smaller pore size. The unsaturated coefficient of hydraulic conductivity of the fine grained soil layer is higher than that of the coarse grained soil layer at high suction values (Rahardjo et al. 2012).

As a result of the interface capillary forces, the water will not be able to infiltrate the coarse grained soil layer. Instead, it will be stored within the fine-grained soil layer (therefore, termed "storage layer") by capillary action (Khire et al. 2000; Bussiere et al. 2003; Yang et al. 2004;

Krisdani et al. 2005; Rahardjo et al. 2009). The stored water will be later drained using one of the capillary barrier drainage techniques discussed further in this chapter.

The differences in the grain-size distribution of fine and coarse grained soils will also have a significant influence on the soil water characteristic curve (SWCC) as well as on the hydraulic properties of the soils. As a result, the soil hydraulic conductivity and the volumetric water content will vary across the interface between those two materials (Yang et al. 2004; Rahardjo et al. 2009). Figures 1.2a and 1.2b (see Chapter 1) show the SWCCs, which give the relationship between the volumetric water content and the coefficient of hydraulic conductivity versus matric suction.

Rahardjo et al. (2009) stated that the complete understanding of the mechanism of operation of the capillary barrier system could be practically achieved by knowledge of the SWCCs, in addition to the hydraulic conductivities for both soils (i.e., the fine and coarse-grained soils).

Flat (horizontal) capillary barriers (from Lu and Likos 2004a)
The principles of how the capillary barriers work is explained by Lu and Likos (2004a) through an equilibrium analysis of hydraulic head within a two layer system under hydrostatic conditions (Figure 2.2). The following two sections are taken from Lu and Likos (2004a) with a few minor changes in text.

"The transition between fine and coarse soil layers is presented as cone shaped pore with radii on one of the sides that corresponds to the average pore sizes of the fine and coarse layers".

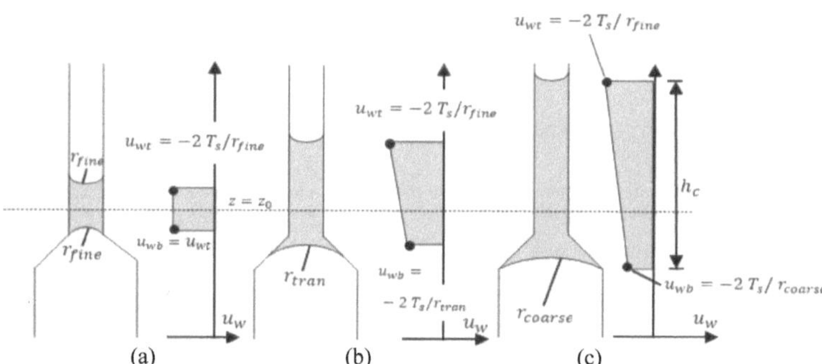

Figure 2.2 Hydrostatic equilibrium of capillary water near interface of fine soil-coarse soil capillary barrier system (a) thin suspended water layer (b) intermediate suspended water layer (c) water layer at threshold of breakthrough (modified after Lu and Likos, 2004a)

In each stage of water accumulation, the hydrostatic or no-flow, equilibrium condition is taking place. This action will lead to a constant head in the vertical direction where the water phase is continuous.

Under the assumption that the solid-liquid contact angle in both soil layers is zero (α=0) and the air pressure is a zero reference value (u_a=0), then mechanical equilibrium at the air-water-solid interface in the fine soil requires the following to be true when the suspended water lens is infinitesimally thin (i.e. very small) as shown in Figure 2.2a, and mathematically is presented in Equation 2.1.

$$u_{wt} = -\frac{2\,T_s}{r_{fine}}$$

(2.1)

where u_{wt} is the pore water pressure at the point near the air-water interface in the fine soil, r_{fine} is a representative pore radius for the fine soil, and T_s is the surface tension.

Similarly, mechanical equilibrium at the air-water-solid interface near the bottom of the water lens leads to:

$$u_{wb} = -\frac{2\,T_s}{r_{fine}}$$

(2.2)

which is equal to the pore pressure near the top since the water lens is infinitesimally thin (i.e., $u_{wt} = u_{wb}$).

With time, the overlying water lens becomes thicker as shown in Figure 2.2b, the total head buildup (i.e., increase) due to gravity requires the pore water to move slightly into the transitional zone between the fine and coarse layers. The pore pressure near the bottom of the lens is greater than that near the top by an amount proportional to the thickness of the water lens and $\rho_w g$. Therefore, at the mechanical equilibrium, the pore pressure near the bottom of the water lens becomes:

$$u_{wb} = -\frac{2\,T_s}{r_{tran}} \qquad (2.3)$$

where:

r_{tran} is the equilibrium radius in the transition zone. As r_{tran} is generally smaller than the representative radius of the coarse soil grains r_{coarse}, but larger than the representative radius of the fine soil grains r_{fine}.

The pore pressure described by Eq. 2.3 is less than the water-entry pressure of the coarse soil layer. The water-entry pressure of the coarse soil (i.e., the pressure at which water begins to enter the coarse soil layer) is mathematically described in Eq. 2.4:

$$u_w = -\frac{2\,T_s}{r_{coarse}} \qquad (2.4)$$

The thickness of the water lens increases due to the infiltration from the ground surface. Accordingly, the pore pressure near the bottom of the water lens increases and the wetting front gradually moves to a new equilibrium position.

When this pressure becomes equal to the water entry pressure of the coarse grained soil layer, the wetting front advances to the position at the end of the transition zone as shown in Figure 2.2c. At this stage, the mechanical equilibrium is at the breakthrough threshold as expressed by Eq. 2.5.

$$h_c \rho_w g = u_{wb} - u_{wt} = \frac{2\,T_s}{r_{fine}} - \frac{2\,T_s}{r_{coarse}} \qquad (2.5)$$

where h_c is the "breakthrough" or the critical head. At this point, the capillary barrier will not function any more and water will enter into the coarse soil layer, as illustrated through Eq. 2.6.

$$h_c \geq \frac{2\,T_s}{\rho_w g r_{fine}} - \frac{2\,T_s}{\rho_w g r_{coarse}} \qquad (2.6)$$

The first term on the right side of the above equation controls the magnitude of the minimum pore water pressure. In contrast, the second term controls the maximum pore water pressure as shown in Figure 2.2c. Accordingly, the larger the difference between the two terms, the more effective the capillary barrier is.

The soil-water characteristic curve (SWCC) is a key parameter that should be taken into consideration during the design of capillary barrier systems. For a perfectly wetting material, the representative pore radius, r_{fine} may be related to the air-entry pressure value, u_b or the parameter, α that is used to model the soil-water characteristic curve.

$$r_{fine} = \frac{2\,T_s}{u_b} = 2\,T_s\,\alpha_{fine} \tag{2.7}$$

Similarly, the representative pore radius for the coarse soil r_{coarse} may be related to the water-entry pressure, u_w, which is suggested to be half of the air-entry pressure value

$$r_{coarse} = \frac{2\,T_s}{u_w} = T_s\,\alpha_{coarse} \tag{2.8}$$

Considering Eqs. 2.7 and 2.8, Eq. 2.6 is rewritten as

$$h_c \geq \frac{1}{\rho_w g \alpha_{fine}} - \frac{2}{\rho_w g \alpha_{coarse}} \tag{2.9}$$

Dipping capillary barriers (inclined or slope capillary barrier systems) (from Lu and Likos 2004a)

As inclined or sloping capillary barrier is more effective than a horizontal capillary barrier system, especially in the case of pre-wetted coarse soil layer, or in the case of continued downward infiltration. It is designed by dipping the interface between two layers (i.e., the fine and coarse soil layers) to force the water to flow along the dipping portion of the interface under the combined forces of capillarity and gravity.

The lateral distance from the highest portion of the dipping barrier to the point of breakthrough is referred to as the diversion width, L. The total flux of water that is diverted is called the diversion capacity, Q. Diversion capacity, Q is equal to qL, where q is the steady infiltration flux.

Steenhuis et al. (1991) proposed a model for the hydraulic conductivity function that includes the air-entry head of the fine soil layer, $h_{a,fine}$ and the water-entry head, $h_{w,coarse}$ of the coarse layer.

$$k = \frac{k_s}{k_s} \exp[\gamma_w \alpha_{fine}(h_m + h_{a,fine})] \qquad \begin{array}{ccc} |h_m| & < & h_{a,fine} \\ |h_m| & \geq & h_{a,fine} \end{array} \qquad (2.10)$$

which allowed to the following solution for diversion width:

$$L \leq \frac{k_s tan\phi}{q} \left[\frac{1}{\gamma_w \alpha_{fine}} + (h_{a,fine} - h_{w,coarse}) \right] \qquad (2.11)$$

The diversion capacity is derived from Eq. 2.11.

$$Q_{max} \leq k_s tan\phi \left[\frac{1}{\gamma_w \alpha_{fine}} + (h_{a,fine} - h_{w,coarse}) \right] \qquad (2.12)$$

The above equations are better suited to the case when the air entry head is nonzero. If the exchangeability between the term $1/\gamma_w\,\alpha$ and the parameter h_a is defined, then the above equations becomes:

$$L \leq \frac{k_s tan\phi}{q} (2h_{a,fine} - h_{w,coarse}) \qquad (2.13)$$

$$Q_{max} \leq k_s tan\phi\, (2h_{a,fine} - h_{w,coarse}) \qquad (2.14)$$

The water-entry head, h_w has also been related to air-entry head, h_a, Bouwer (1966) suggested that the air-entry head is taken as twice the water-entry head, while Walter et al. (2000) suggested that the air-entry head is equal to the water-entry head. For $h_w = {h_a}/{2}$, Eqs. 2.13 and 2.14 become:

$$L \leq \frac{k_s tan\phi}{2q} (4h_{a,fine} - h_{a,coarse})$$

(2.15)

$$Q_{max} \leq \frac{k_s tan\phi}{2} (4h_{a,fine} - h_{a,coarse})$$

(2.16)

For $h_w = h_a$, Eqs. 13 and 14 become:

$$L \leq \frac{k_s tan\phi}{q} (2h_{a,fine} - h_{a,coarse})$$

(2.17)

$$Q_{max} \leq k_s tan\phi (2h_{a,fine} - h_{a,coarse})$$

(2.18)

For an optimum design of the capillary barrier, the air-entry head of the coarse soil should be much smaller than that of the fine soil and the maximum diversion width becomes:

$$L_{max} \leq \frac{k_s h_{a,fine} tan\phi}{q}$$

(2.19)

For design purposes, the efficiency of a capillary barrier may be defined as $\omega = {L}/{L_{max}}$. Combining Eqs. 2.17 and 2.19 lead to

$$\frac{h_{a,fine}}{h_{a,coarse}} = \frac{1}{2 - 2\omega}$$

(2.20)

A less conservative expression for the ratio is derived by combining Eq. 2.15 and Eq. 2.19:

$$\frac{h_{a,fine}}{h_{a,coarse}} = \frac{1}{4 - 4\omega} \qquad (2.21)$$

Both Eqs. 2.20 and 2.21 provide practical design guidelines for selection of soil to be used in capillary barrier systems.

2.3 Soil- Water Storage Capacity

The soil-water storage capacity is the amount of water stored within the fine grained soil layer of a capillary barrier system, which acts as a moisture retaining layer (Bussiere et al. 2000; Krisdani et al. 2005). Rahardjo et al. (2009) have concluded that the water storage capacity of the fine grained soil layer is the most important factor, which can determine the applicability of the capillary barrier system as a slope stabilization technique. Therefore, increasing the water storage capacity is an objective for the geotechnical engineers that should be considered in the design of such systems (Khire et al. 2000). Many factors could affect the efficiency of the moisture storage layer, such as slope geometry (for slope-capillary barrier systems), the thickness of the fine grained soil layer, and the material properties (Bussiere et al. 2000).

Recovery of Soil Water Storage capacity

The water in the storage (fine-grained soil) layer can be removed using various methods, such as evaporation (natural process), and lateral drainage (Yang et al. 2004; Rahardjo et al. 2009).

The recovery of soil-water storage capacity is one of the most important variables to be taken into account in the design of capillary barrier systems, especially in high precipitation regions.

The recovery process should take place rapidly so that the system may function efficiently during dry periods and be able to store additional amounts of the water during periods of high infiltration and rainfall. The water, entering the system, should be prevented from reaching the interface between the fine grained soil layer and coarse grained soil layer (Krisdani et al. 2005; Rahardjo et al. 2009).

2.4 Design of capillary barriers

In simplest terms, the main purpose of the design of a capillary barrier is to provide the thickness of the fine and coarse grained soil layers that will eliminate or significantly reduce the infiltration of water into the waste. However, considering the difficulties involved in the accurate determination of the influential factors, the design of capillary barriers becomes not that simple. Khire et al. 2000 describe the following steps in a design procedure.

(a) Determine the source of soils to be used for the fine and coarse grained soil layers

It is economical to use soils if they are available locally. Clean sand or gravel should be used for the coarse grained soil layer.

(b) Determine the critical meteorological period

The design should be based on site-specific meteorological and hydrological conditions. Critical values used for precipitation (including snow melt) should be based on both short term extreme events and the cumulative values of precipitation for the complete length of the design life. Considering the possible changes in the climate of the earth, this step should include possible modifications to the existing meteorological data.

(c) Determine the hydraulic properties of soils to be used in the capillary barrier

The list of parameters include SWCC, soil water storage capacity, and hydraulic conductivity.

(d) Estimate the thickness of the coarse and fine grained soil layers

The thickness of the coarse grained soil layer can be chosen as 30 cm considering the placement of the soil in the field conditions. An estimate of the thickness of the fine grained surface layer can be made by using a method such as the one proposed by Stormont and Morris (1998) or Ross (1990). Actually, this step provides a thickness for both coarse and fine grained soil layers so that a numerical analysis can be initiated as described in the next step.

(e) Adjust thickness

The thickness of the fine grained soil layer can be adjusted by conducting water balance simulations with one of the numerical models capable of seepage analysis in unsaturated soils.

(f) Account for "Other Factors"

The other factors that are not accounted for in the steps (a) to (e) are water erosion, wind erosion, and desiccation cracking.

(g) Include the requirements of the applicable codes and regulations

2.5 Codes and Regulations Related to Soil Covers over Waste in Landfills

In Canada, provincial governments publish codes and regulations for the disposal of municipal solid waste within their jurisdictions. For example, the Code of Practice for Landfills (2010) by the Alberta Government Environmental Protection Branch states the following. *"The final cover system shall include a barrier layer of 0.60 metres of earthen material with a maximum permeability of 1 x 10^{-7} metres/second"*. It should be noted that all provincial codes and regulations fall under the Canadian Environmental Protection Act (1999). The enforcement of the regulations and the publication of detailed documents are left to the provinces.

In Ontario, the Ministry of the Environment has the authority to ensure safe disposal of waste. A guideline on regulations is given in the publication Landfill Standards (2012). *Regulation 232/98* contains all requirements for the design, operation, closure and post-closure care of municipal (i.e. non-hazardous) waste landfilling sites. Some of the regulatory requirements for the final cover are listed below.

1. *The owner and the operator of a landfilling site shall ensure that the following materials are applied to the waste fill zone as final cover, from bottom to top:*

    1. *A minimum of 0.6 metres of cover material.*

    2. *A minimum of 0.15 metres of topsoil or other material approved by the Director as able to sustain plant growth.*

    3. *A vegetative cover consisting of vegetation that is suited to local conditions and that is capable with minimal care of providing vigorous, plentiful cover not later than its third growing season.*

2. *The owner and the operator of a landfilling site shall ensure that the final cover is designed so that,*
    a. *the infiltration rate through the final cover is in accordance with the design for the site respecting ground water protection prepared under section 10;*

    b. *any existing or anticipated facilities for the control, collection, use or discharge of landfill gas are accommodated; and*

    c. *the requirements for the end use of the site, as described in the site design report prepared under section 6 and the closure report prepared under section 31, are met.*

Regulation 232/98 also states that the final cover is aesthetically pleasing, controls infiltration, and is suitable for the end use planned for the site. The final cover should also be compatible with any gas control needs for the site. Infiltration rates are also regulated to ensure the service life of the engineered facility exceeds the life span of the contaminants.

Regulation 232/98 includes statements about the slope angles of the landfill. Steep slopes promote surface water runoff, but increase erosion potential. More gradual slopes reduce runoff and erosion, but increase infiltration and the potential for ponding of surface water. Each of these factors should be considered in the design of the site and choice of final contours.

Regulation 232/98 sets limits on maximum and minimum final slopes, but allows alternative designs where appropriate. This requirement is given in Section 30 of the Regulation and is as follows:

Section 30

1. *The owner and the operator of a landfilling site shall ensure that the final slopes above grade within the waste fill zone at the time of site closure do not exceed one unit vertical to four units horizontal and are not less than one unit vertical to 20 units horizontal.*
2. *Subsection (1) does not apply if a written report has been prepared that confirms that an alternative design for the final slopes is acceptable, having regard to the slope stability of the deposited waste and final cover, the potential for erosion of the final cover, the proposed end use of the site and the infiltration requirements for ground water protection.*

British Columbia Ministry of Environment (2016) published a document called "Landfill Criteria for Municipal Solid Waste". That document includes regulations relevant to soil covers over municipal solid waste. The following paragraph is a copy of the regulation related to soil cover.

"The minimum final cover shall consist of a barrier layer, providing a maximum hydraulic conductivity of 1 x 10-5 cm/sec for landfill sites located in arid and semi-arid regions and 1 x 10-7 cm/sec for landfill sites located in non-arid regions. Though arid and non-arid regions are not only characterised by annual precipitation levels, as a general guidance, areas with less than 500 mm of annual precipitation can be considered as arid and semi-arid. The final cover barrier layer shall have a minimum compacted thickness of 0.6 m measured perpendicular to the slope

with a minimum 0.15 m topsoil layer capable of establishment and sustained growth of the vegetative cover."

The minimum final cover requirements are illustrated in Figure 2.3.

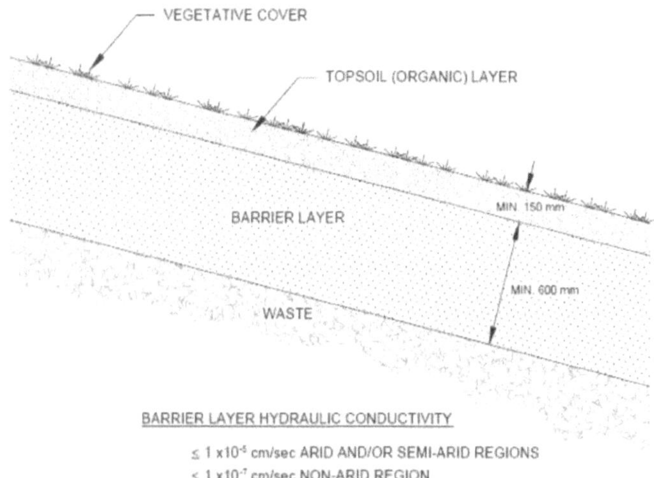

Figure 2.3 Final cover requirements stated in the regulation of BC Ministry of Environment

In the case of the disposal of radioactive waste, the federal government in Canada has strict regulations in all aspects of the disposal.

In the United States, Environmental Protection Agency (EPA) plays a major role in the preparation of regulations/guidelines for the disposal of any kind of waste. Detailed information about how to build covers for solid waste and hazardous waste can be found in publications such as EPA (U.S. Environmental Protection Agency), 1989a, EPA (U.S. Environmental Protection Agency), 2004, and ITRC (Interstate Technology & Regulatory Council), 2003.

2.6 Soil physical properties

2.6.1 Soil water charateristic curve

The soil-water characteristic curve (SWCC) presents the relationship between the water content of the soil (volumetric water content, gravametric water content, or degree of saturation) and suction (Fredlund and Xing. 1994; Vanapalli et al. 2002; Zhai and Rahardjo 2012).

Figure 2.4 shows a soil water characteristic curve with its main features.

Soil-water characteristic curve provides the basic knowledge, which is required to describe the mechanical behavior of the unsaturated soils (Fredlund and Xing 1994; Zhai and Rahardjo. 2012).

Suction has two components namely, matric suction and osmotic suction. Both matric and osmotic suctions make-up the total suction (total suction = matric suction + osmotic suction) (Fredlund et al. 2012).

$$\Psi = (u_{a-} u_w) + \pi \tag{2.22}$$

Where Ψ is the total suction (kPa), u_a is the pore air pressure, u_w is the pore ware pressure, π is the osmotic suction (kPa).

Matric suction is defined as the capillary pressure of the soil, which is equal to the pore-air pressure minus pore-water pressure (u_a-u_w). The osmotic suction, in turn, is related to the salts content in the soil structure (Fredlund et al. 2011).

Fredlund and Xing (1994) defined the air-entry value of the soil as the matric suction at which air starts to enter the largest pores in the soil during the drying process. Yang et al. (2004) defined the residual water content as the water content at which a discontinuous condition appears for the water phase.

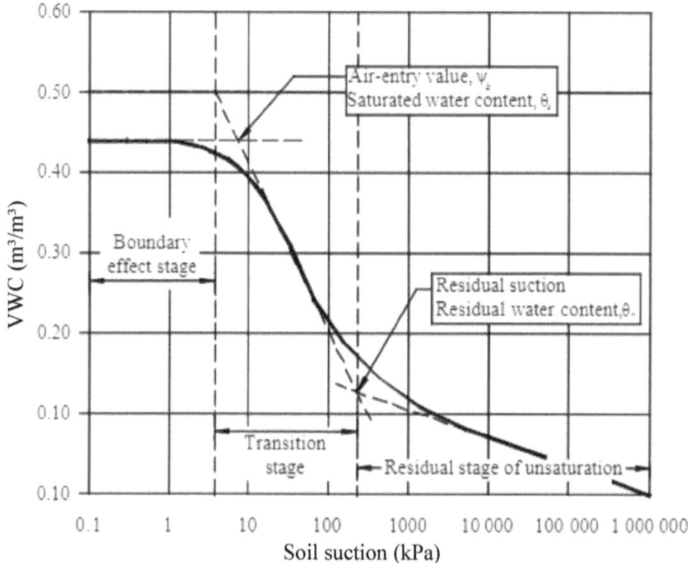

Figure 2.4 The typical soil water characteristic curve (Vanapalli et al. 2002)

Vanapalli et al. (2002) have defined the process of soil desaturation (Figure 2.5). Referring to Figure 2.5, the boundary stage effect is observed when the soil is fully saturated. During the transition stage (desaturation zone), the soil begins to desaturate, that is, the water content of the soil is reduced and the water menisci at contact points with soil grains are discontinuous.

This stage is divided into two substages namely, primary and secondary transition stages. In the residual stage, no significant change takes place in the water content but the matric suction continues to increase. The water content becomes equal to zero at the end of the residual stage. The movement of water through soil in the residual stage is mostly in vapor phase.

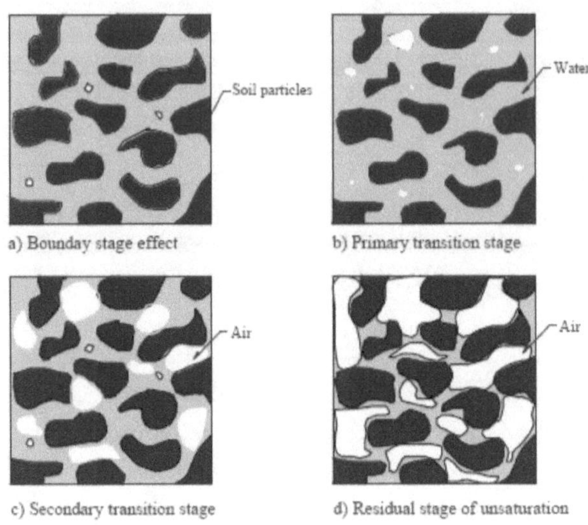

Figure 2.5 The different stages of desaturation (Vanapalli et al. 2002)

Van Genuchten (1980) has proposed a mathematical expression to estimate the soil-water characteristic curve (SWCC) as expressed in the following equation;

$$\theta = \theta r + \frac{(\theta s - \theta r)}{(1 + (\alpha \Psi)^n)^m} \tag{2.23}$$

Fredlund and Xing (1994) have developed a model to estimate the soil-water characteristic curve (SWCC) as expressed in the following equation;

$$\theta = \theta s \left[\frac{1}{\ln\left[e + (\Psi/\alpha)^n\right]} \right]^m \tag{2.24}$$

In Eqs. 2.23 and 2.24, θ is the volumetric water content (m³/m³), θ_r is the residual volumetric water content (m³/m³), θ_s is the saturated volumetric water content (m³/m³), Ψ is the soil matric suction (kPa), α, n, and m are the curve fitting parameters.

Eqs. 2.23 and 2.24 are the backbone (the fundamental equations) in the literature used to estimate the SWCC.

2.6.2 Soil-water evaporation

The evaporation from the surface of the soil has been studied by various researchers, such as Wilson et al. (1995; 1997).

Wilson et al. (1995) have described the soil-water evaporation as the flow of moisture between the soil surface and the atmosphere. They indicated that the actual rate of evaporation is a function of the total suction and the flow of moisture takes place in both liquid and gas phases.

The actual rate of evaporation (AE) from the soil can be affected by the atmospheric demand called the potential evaporation (PE) and the water content of the soil (Wilson et al. 1995).

Potential evaporation (PE) is the maximum evaporation rate. According to the World Meteorological Organization, PE is defined as "The quantity of water vapor which could be emitted by a surface of pure water per unit surface area and unit time under the existing atmospheric conditions" (Wilson et al. 1997).

Using the mass transfer equation (Dalton's equation) the potential evaporation from free water surface can be determined using Eq. 2.25 (Wilson et al. 1995; 1997), where the evaporation from free water surface is considered to be the same as the evaporation from the saturated soil (wet). It was found that the PE was affected by several factors such as the condition of the atmosphere, and the temperature of the surface (Wilson et al. 1997).

$$PE = f(u)(e_s - e_a) \tag{2.25}$$

where $f(u)$ is transmission function which depends on the mixing characteristics of the air above the evaporation surface (estimated empirically); e_s is the saturation vapour pressure of the water at the temperature of the surface; e_a is the vapour pressure of the air in the atmosphere above the water surface.

Wilson et al. (1995; 1997) proposed that the ratio of the actual evaporation to the potential evaporation (normalized soil evaporation) is equal to unity (i.e., 1), which implies that both values of evaporation are equal. This theory is applicable for sandy soils as well as for clayey as long as the water content of the soil is high (saturated or near saturated level). The factors which can influence the AE/PE ratio are evident from Figure 2.6 that displays the relationship between the water content, soil texture (sand or clay), and the evaporation rate (slow or fast).

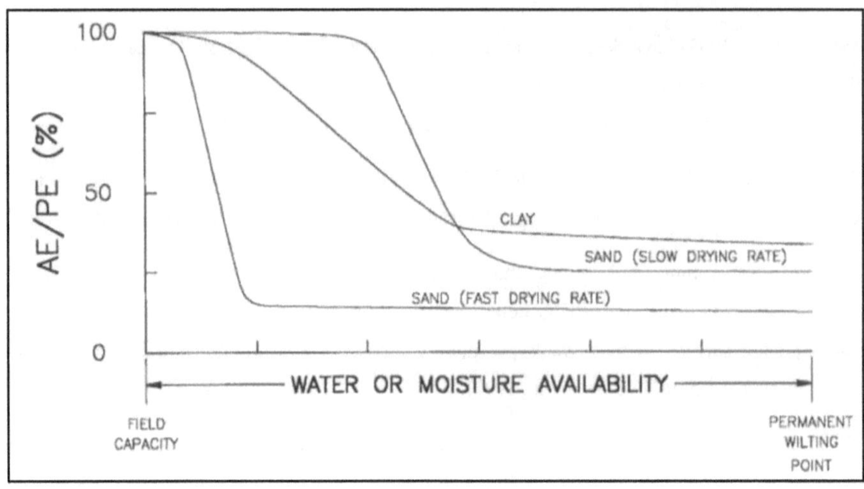

Figure 2.6 The relationship between AE/PE ratio and water content (Wilson et al. 1997)

The relative humidity was found to have a significant effect on the evaporation rate. As the evaporation increases, the humidity will also increase. Wilson et al. (1995; 1997) proposed that the relative humidity is a function of total suction. As the suction increases, the relative humidity decreases (Wilson et al. 1995). It can be seen that the value of the relative humidity and the soil actual evaporation depend on the total suction of the soil, in other words the relative humidity and the actual evaporation are a function of the total suction. If the value of the total suction reaches 3000 kPa (the evaporation-rate reduction point), then the relative humidity and the actual evaporation will start decreasing as shown in Figure 2.7.

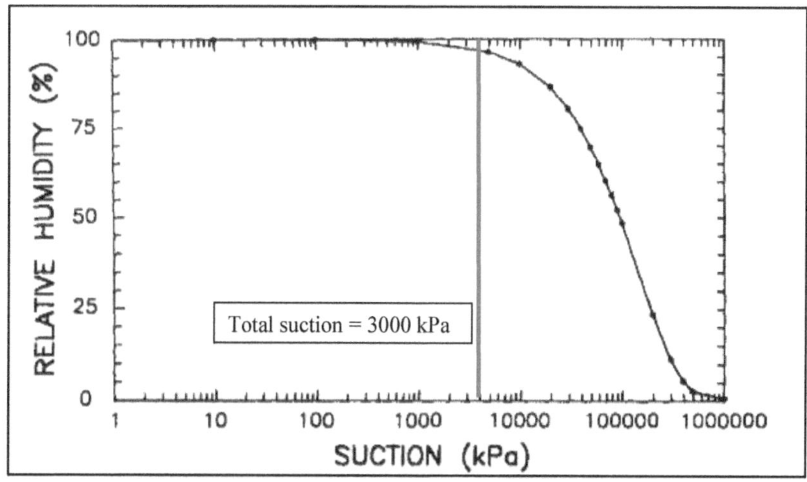

Figure 2.7 Relative humidity versus suction based on humidity-suction equation (Wilson et al. 1995)

The relative humidity in the soil is expressed mathematically as a function of the total suction (Wilson et al.1995) as follows,

$$P_v = exp\left(\frac{\Psi g\, W_v}{RT}\right) P_{sv} \qquad (2.26)$$

where P_v is the actual vapor pressure within unsaturated soil voids; exp is exponential function; Ψ is total suction; g is the acceleration of gravity (m/s^2); W_v is the molecular mass of water (0.018 kg/mol); R is the universal gas constant (8.314 J/ (mol·K)); T is the temperature; P_{sv} is the saturation vapour pressure at temperature T. The above equation indicates that the actual vapour pressure in the soil is equal to the product of the relative humidity and the saturated vapour pressure (P_{sv}).

The above equation can be rewritten in terms of the relative humidity as shown in Equation 2.27 (Wilson et al. 1995).

$$hr = (P_v/P_{sv})$$

(2.27)

By substitution of Equation 2.26 into Equation 2.27, the final form for the relative humidity equation as function of suction is written as,

$$hr = exp\left(\frac{\Psi g\, W_v}{RT}\right)$$

(2.28)

where hr is the relative humidity.

Based on the relationship between the relative humidity and soil evaporation rate, it was found that the evaporative fluxes begin to decline when the actual vapour pressure in the soil becomes equal to the actual pressure in the air (Wilson et al. 1995). It was concluded that the vapour pressure performance will change from saturated to unsaturated as total suction is increased in excess of 3000 kPa. On the other side, the evaporation will decrease when the relative humidity is decreased as a function of suction.

Figure 2.8 shows that the total suction has a significant effect on the evaporation process, while the evaporation ratio (AE/PE) starts decreasing beyond a total suction value of above 3000 kPa and becomes equal to zero at a suction of 100,000 kPa. The total suction is affected by heating of the soil (i.e., it decreases with the increase in temperature) and that will ensure the evaporation process of the soil.

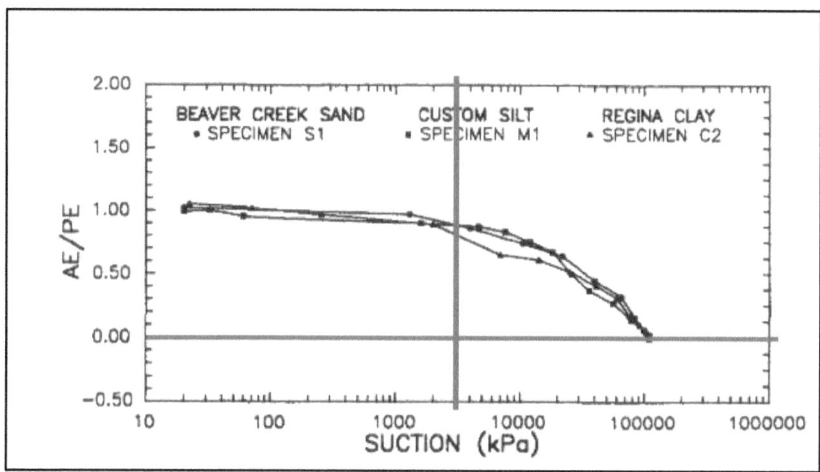

Figure 2.8 AE/PE ratio versus total suction for three types of soil (Wilson et al. 1995)

2.7 Dry capillary barrier systems

The concept of dry capillary barrier systems has been discussed by a large number of researchers (e.g., Stormont et al. 1994; Stormont et al. 1998; Morris et al. 1999; Morris 1999; Albrecht and Benson, 2002). A dry capillary barrier is defined as a system where dry air (atmospheric air) is circulated within the coarse soil layer of a capillary barrier. Such systems are commonly used at landfill sites as a surface cover over waste material. The purpose of the dry capillary barrier system is to decrease the water content of the soil in the coarse grained soil layer by evaporation especially near its interface with the fine grained soil layer. The evaporation rate is a function of the air flow rate among other factors such as temperature and humidity in the air. As the dry air goes through the soil, the water is removed as a vapor phase. Table 2.1 shows a summary of the characteristics of a dry barrier system.

Table 2.1 Summary of the characteristics of dry barrier systems summary (Stormont et al. 1998)

Approach	Application	Basic concept	Method of air induction	Mathematical Model
Dry barrier system	Landfill Monitoring (detect the changes in the humidity rate in the coarse layer)	Air flow is applied to reduce moisture in the coarse layer and this, prevent or delay percolation. Technique also applied for early detection of percolation in the capillary barrier systems.	(Bulk air) Active system (Blower), Passive system (wind-powered chimney)	$\dfrac{dR}{dt} = (a_{out} - a_{in})\dfrac{dv}{dt}$

One of the advantages of this technique is the restoration (i.e., recovery) of any capillary barrier system which was previously flooded due to excess amount of rain. Recovery of the system is possible by using the evaporation process that occurs when airflow is circulated within the coarse layer (Stormont et al. 1998). The airflow also increases the water storage capacity of the fine soil layer (Morris et al. 1999).

Many researchers (e.g., Stormont et al. 1998; Morris et al. 1999; Morris 1999) have indicated that the dry air plays an important role in the evaporation process, as it has lower moisture content (lower humidity) than the unsaturated soil, and thus will dry effectively the latter.

Unlike the pore gas in equilibrium within the unsaturated soil, the atmospheric air has a lower water vapor. Dry air is circulated to remove moisture from the soil. In the case when the coarse layer of the system is initially saturated, it should be dried first to ensure proper operation of the barrier system. That can be achieved with one of the drainage techniques that creates paths for air to flow. Hence, such a technique should be combined with the dry barrier system to ensure its effectiveness.

Albrecht and Benson (2002) studied dry barrier systems using the passive dry barrier shown in Figure 2.9 They utilized the wind velocity to induce air flow in the coarse soil layer and proposed the following hypothesis for the operation of the passive dry barrier system. The drop in pressure in the coarse grain soil occurs due to the increase of air velocity and leads to development of a driving force for the moisture to move from the fine grained soil layer to the interface between fine-grained and the coarse-grained soil layer. Bernoulli's law can be used to explain this phenomenon, where the moisture moves from a high pressure toward a low pressure region.

The difference in the pressures at both ends of the barrier (i.e., the inlet and the outlet pipe) induces gradient in the air pressure within the coarse soil layer that drives the air flow in that layer (Figure 2.9).

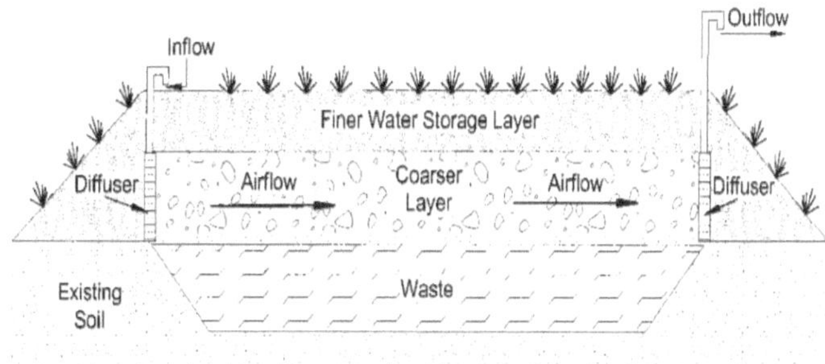

Figure 2.9 The passive barrier system (from Albrecht and Benson 2002)

Stormont et al. (1994) concluded that when a water pressure gradient develops in the capillary barrier system due to a decrease in the hydraulic head caused by air flow, water will be driven toward the interface between the fine-grained and the coarse-grained soil layer and there it will be evaporated through the air flow, as illustrated in Figure 2.10.

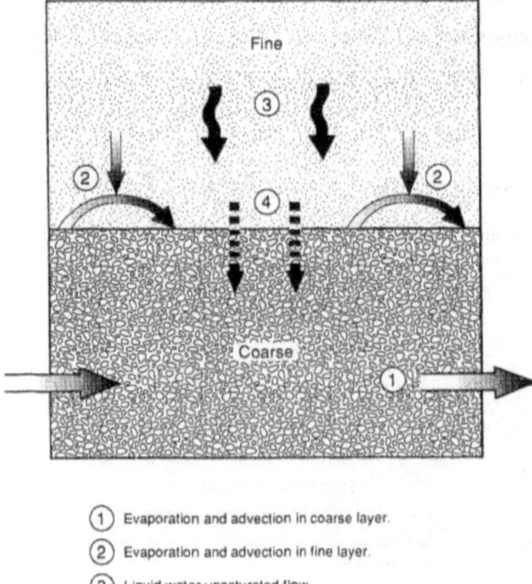

Figure 2.10 Water movement in a capillary barrier system due to air flow (Stormont et al. 1994)

Stormont et al. (1998) used a network of pipes in a landfill cover system to introduce and distribute air within the coarse layer (the air-permeable layer), as shown in Figure 2.11 The system has two components: a pipeline for pressurized air supply (pressure line), and a vacuum line to remove air enriched with water vapor from the system. The air has advection motion in horizontal direction. The purpose of the air flow is to prevent water from reaching the coarse layer, and to maximize the soil's water storage capacity.

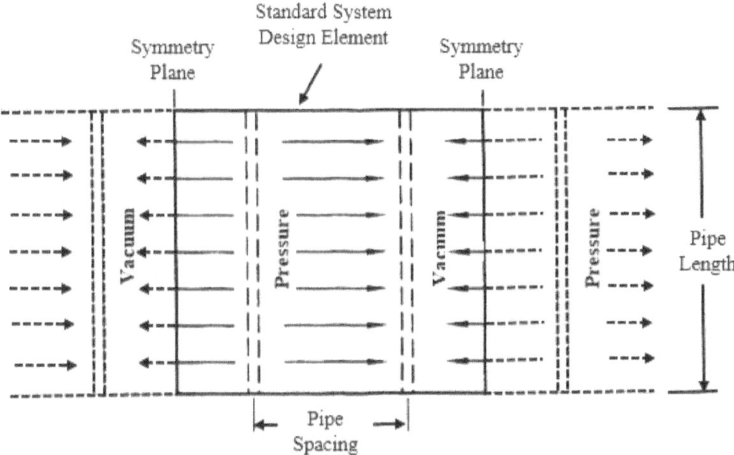

Figure 2.11 Cross section of the dry barrier system using a pipe network (Stormont et al. 1994)

Stormont et al. (1998) proposed a mathematical model given by Equation 2.29 to calculate the net water removal rate from a system that uses the dry barrier technique.

$$\frac{dR}{dt} = (a_{out} - a_{in})\frac{dv}{dt} \tag{2.29}$$

where dR/dt is the net water removal rate (g/s); a_{out} is absolute humidity of the air leaving the system; a_{in} is absolute humidity of the air entering the system; dv/dt is volumetric air flow rate (m^3/s).

Figure 2.12 shows the removed water (as a mass fraction and in cm) versus time for different applied air pressure gradients.

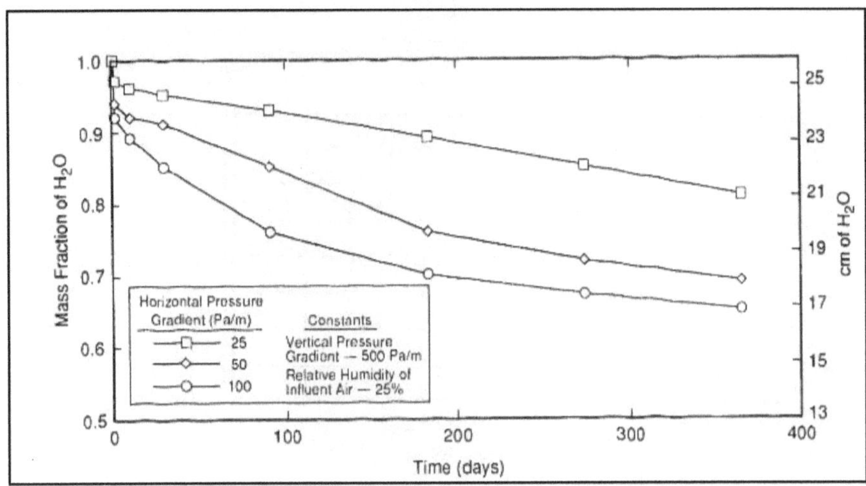

Figure 2.12 Water removal from the dry barrier versus time (Stormont et al. 1994)

Morris (1999) demonstrated that using bulk atmospheric air in the capillary barrier system has a significant effect on the evaporation rate. The effect of circulated air on the moisture removal rate has been studied in the past by several researchers (e.g., Stormont et al. 1994; Stormont et al. 1998; Morris et al. 1999; Morris 1999).

A capillary barrier system, where the air was circulated in the coarse layer to restore its dry condition, was proposed for monitoring a waste containment facility liner (Stormont et al.1998). The proposed system also had the ability to detect leaks by monitoring the changes in the humidity of the pore gas in the coarse soil layer (Stormont et al.1998).

2.8 Methods for humidity control

Humidity control is a big challenge and main objective in greenhouses (Agricultural field). Whereas, high humidity can cause a fungal infection (due to condensation process) and irregular transpiration process.

Air temperature and moisture content in the air can be considered key factors in controlling the humidity of air. Dehumidification process is the process which leads to reduction of moisture in the air (by replacing the moist air with dry air). It aims to prevent reaching the dew point at which the air accumulates close to 100 % humidity, which starts condensing as soon as the air temperature drops below the dew point temperature. Any variation in the temperature between soil surface and the atmosphere causes problems. Keeping both temperatures equal to each other as much as possible especially at high relative humidity of air is therefore important (Agriculture facts sheet 1994; Humidity/Moisture Handbook 1999).

In the literature, there are many research areas in different disciplines, which employ dehumidification in their applications, such as air conditioning engineering. One study of air conditioning process showed that efficient ventilation can be achieved with a combination of cooling and dehumidification processes replacing the hot and moist air with cold and dry air using an air conditioning system (Ayad Mustafa 2011).

Dehumidification through a pipe buried in the soil mass has been previously applied and discussed in the literature (Boulard et al. 1989; Hollmuller and Lachal 2000, 2005). These studies focus on cooling and preheating of civil structures using buried pipe systems, powered by renewable energy sources. The authors used the principles of agricultural green houses to force atmospheric air through buried pipes in a process termed "hypocaust". In their model, they studied the heat and mass transfer processes between the air and the pipe on one hand, and between the pipe and the soil, on the other. As illustrated in Figure 2.13, the water vapor exchange occurs between the saturated air layer (the interface layer immediately above the water surface) and the air flow within the pipe. Studying the transfer (heat diffusion, exchange) between air, the border of the pipe (pipe wall), and the surrounding soil was the objective of the authors.

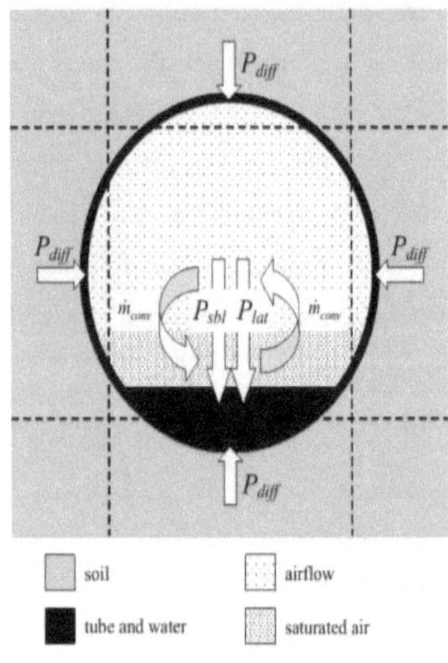

Figure 2.13 Mass and energy at the pipe nodes (Hollmuller and Lachal, 2005)

They did not consider, however, the relationship between the air and the soil, i.e., the soil vapor diffusion due to heat and air flow. The soil's thermodynamic performance and drying due to air temperature were not the point of interest in their research (e.g., Boulard et al. 1989; Hollmuller and Lachal 2000, 2005). Instead, they studied the effect of availability of water in the system due to condensation into the pipe in addition to the air flow through the pipe, and proposed a mathematical model, accounting for the above factors. The model simulates the mass and energy (heat) exchange between the air inside the pipe and the buried pipe itself, i.e., the latent and sensible heat transfer within the pipe.

Agricultural engineering is the other major discipline which uses the dehumidification process to avoid condensation as a result of the high air humidity in greenhouses; this environmental process has a negative effect on the plants.

Vox et al. (2010) and Kittas et al. (2012) postulated that a reduction of air humidity and condensation prevention inside a greenhouse can be achieved through different techniques, such as heating the air inside the greenhouse. Floor heating is an example of a heating method, which is based on heating the air using solar radiation during daytime and then using this hot air during nighttime to heat the surrounding air inside the greenhouse. The hot air flows inside a network of pipes buried under the soil to keep the humidity of air low and, at the same time, to prevent condensation.

Another source also stated the importance of taking into account the dew point temperature (t_d), which has a significant effect on the condensation process. It is necessary to keep the air temperature above t_d at all times to prevent condensation or alternatively, to dry the air by replacing the air with high relative humidity (RH) with dry air (Humidity/Moisture handbook 1999).

Boulard et al. (1989a and 1989b) studied heat and mass transfer in buried pipe systems by simulating an underground heat storage system in a greenhouse using a pipe (PVC drain pipe) buried within the greenhouse soil (storage medium). These authors considered the influence of various factors in their experiments including: flow rate, wind speed, global radiation, and the soil moisture content. They developed a model, which can be used to examine the effect of various factors related to the design of buried pipe systems. They considered the heat and mass transfer (a) between air and pipe, and (b) between pipe and soil. Figure 2.14, which shows the cross section of the pipe and its dimensions, is taken from their publication. As it can be seen in the figure, there are perforations in the pipe wall for condensed water drainage. They concluded that the evaporation in the system can be from the condensed water inside the pipe or from the contact area between the perforations and the soil (the latter was not studied). They indicated that the evaporation surface between pipe (holes) and soil amount to 1% of the pipe surface.

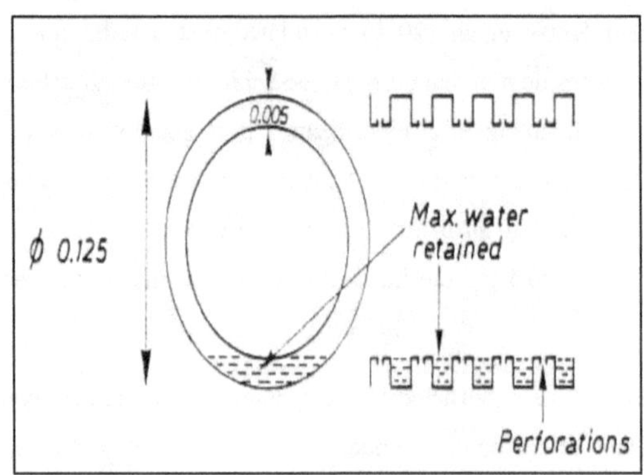

Figure 2.14 The pipe cross section (from Boulard et al. (1989a, and b)

In his doctoral work, Lindblom (2012) studied the Condensation Irrigation (CI) which is a relatively new subsurface irrigation method. In this irrigation method, horizontally placed perforated drainage pipes are installed in the ground below the roots of the vegetation. Heated humid air is forced through the buried drainage pipes. While passing through the pipes, the warm humid air is cooled by the ground and vapour precipitates inside the pipes. The holes in the pipes allow the precipitated water to percolate into the soil and thus, irrigation is achieved. In addition, some of the warm humid air also infiltrates the soil surrounding the pipes through the perforations. As a result, irrigation is increased by vapour condensation in the cooler ground.

Lindblom (2012) carried out both laboratory scale experiments and numerical analyses. Figure 2.15 shows his experimental setup.

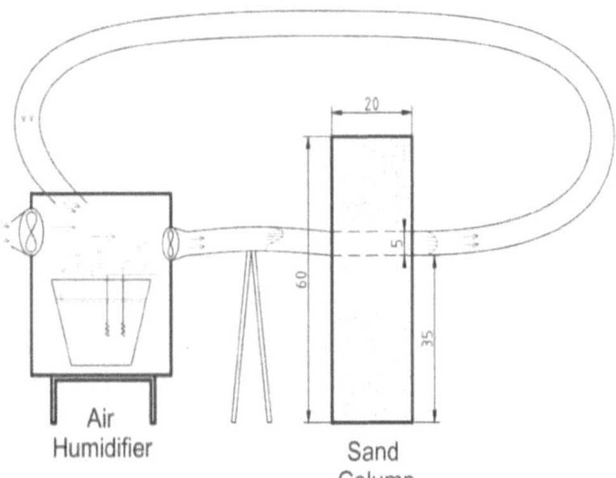

Figure 2.15 Condensation irrigation simulation system (Lindblom 2012, Lindblom and Nordell, 2012) The units are in centimeters

The lab scale model consisted of an air humidifier, a sand column, and a perforated pipe going through the sand column. The dimensions of the sand column were: height = 60 cm, length = 40 cm, and width = 20 cm. The diameter of the pipe was 5 cm. A total of six perforations were made along the pipe. The perforations were 2 mm in diameter and 2 cm apart. The sand column was filled with dry sand before the start of the experiment. Thermocouples were used to measure temperature. Air flow properties were measured inside the pipe, at the entrance and exit of the sand column. These laboratory experiments showed that the air flow through the perforations had a significant effect on the amount of moisture transferred to the soil in this laboratory model of condensation irrigation system. For the numerical simulations, a two dimensional finite element (FE) program was developed using MATLAB. Using this code, which was called CI2D, it was possible to simulate coupled processes involving gas flow, liquid flow and heat transfer in the soil-pipe system. The FE simulations also included water evaporation and condensation. As a simplification, it was assumed that the difference between the wetting and drying portions of the SWCC was negligible, although theoretically, the wetting curve can be much lower than the drying curve.

The following figure shows the moisture content and temperature profiles after 3 days of irrigation.

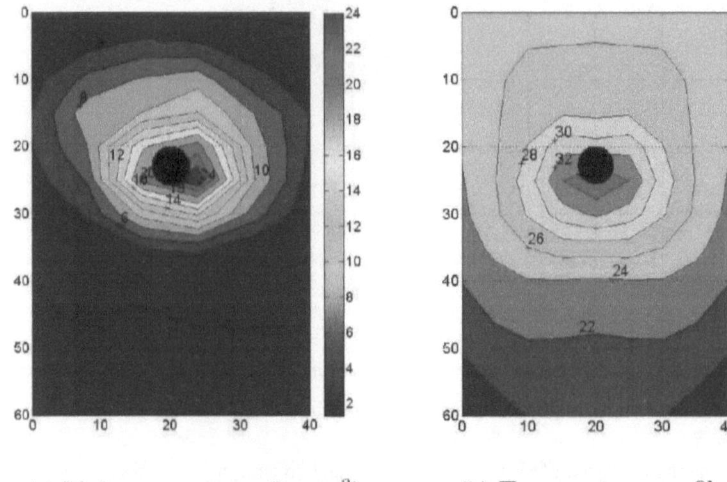

(a) Moisture content (kg m^{-3})   (b) Temperature profile (°C)

Figure 2.16 Moisture content and temperature profiles after 3 days of irrigation (after Lindblom and Nordell, 2012)

Numerical analyses by Lindblom (2007, 2012) showed that the air flow temperature had the greatest impact on the irrigation yield, that was a 20% increase in air flow temperature resulted in a 90% increase in irrigation. Increasing the inlet air flow velocity or humidity 20% resulted in about 10% and 30% more irrigation, respectively.

2.9 Drainage techniques for capillary barrier systems

2.9.1 Capillary barrier systems for protection of slopes

A capillary barrier system used on a slope provides protection against slope instability by preventing or reducing water infiltration into the native soil below the capillary barrier. In this application of the capillary barriers, the water infiltrating the first layer of the barrier moves vertically first, then moves parallel to the slope angle near the interface between the fine and coarse-grained soil layers. The water can then be removed from the system using an external drainage system. For a long slope, the infiltrating water must be collected at some suitable intervals using drainage pipes. In general, the slope geometry at the capillary barrier system operates as a drainage system and the water can flow out easily. However, this technique has a limited dewatering capacity (diversion capacity), which can be affected by numerous factors, such as the thickness of layers, materials properties, slope angle, etc. The diversion capacity can be expressed as the maximum amount (mass) of water which can be diverted by the system, and it depends on the effective length of the barrier system (Bussiere et al. 2003). These authors have indicated that percolation at a particular point within the system occurs when the matric suction at the interface between the fine-grained and the coarse grained soil layers reaches the water entry value (WEV) of the coarse-grained soil layer. This particular point is termed "down-dip limit" (DDL) point. The other important component of a slope capillary barrier system is the effective length of the capillary barrier (Lcb) defined as the distance between the highest point in the slope and the DDL point. Another Lcb definition is the distance which allows water to be diverted before percolation (Morris and Stormont 1999) as shown in the following figure (Note: Lcb and L_D are identical).

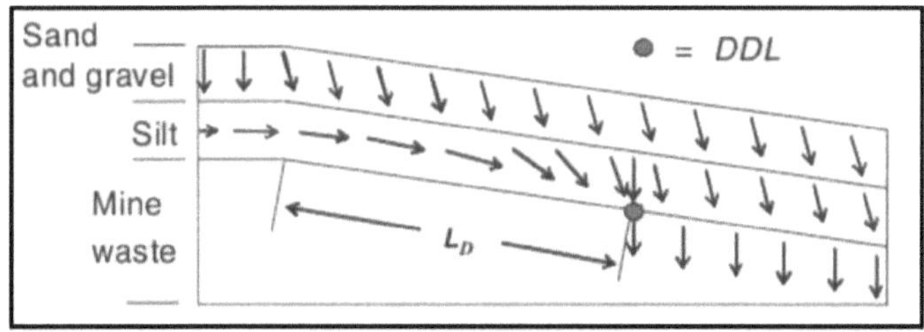

Figure 2.17 Down-dip limit (DDL) point and effective length of a slope capillary barrier (Lcb) (Aubertin et al. 2009)

Bussiere et al. (2003) reached the conclusion that water can be diverted and it would flow along the interface between the fine grained and the coarse grained soil layer down to DDL point. Past the DDL point, water will percolate at the down slope region with increasing moisture content. The location of the DDL point would vary and it depends on several factors, for example, the precipitation rate. The effective length and the diversion capacity would be affected by the saturated hydraulic conductivity (K-sat) of the fine grained soil, slope angle, and the type of soil. In summary, the efficiency of the slope capillary barrier system and its ability to drain water depends on the diversion capacity of the system and requires attention to the various design elements.

As shown in Figure 2.18, the slope-capillary system operation involves several processes: precipitation, infiltration, lateral diversion, breakthrough (percolation), and drying through evaporation (Krisdani et al. 2005; Rahardjo et al. 2009).

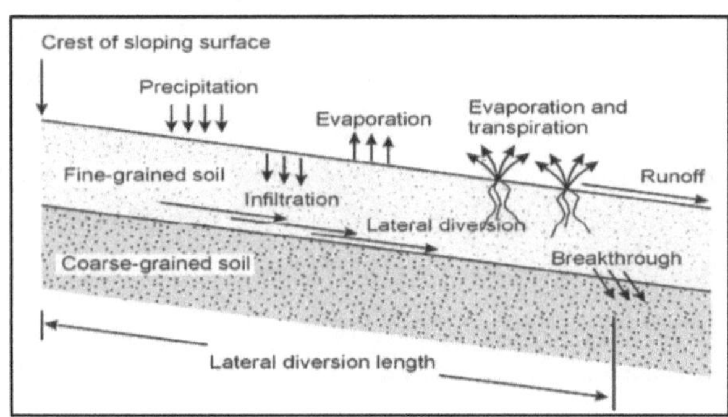

Figure 2.18 The Slope-capillary barrier system with associated processes (Rahardjo et al. 2009)

2.9.2. The unsaturated drainage layer (UDL) or the transport layer

The use of an unsaturated drainage layer (UDL) can improve the performance of the capillary barrier system and increase the water diversion capacity. This technique involves placing a soil layer between the fine and coarse grained soil layers to increase the diversion capacity of the system (Figure 2.19). The water is diverted laterally along the UDL layer, and as a result, it is prevented from reaching the coarse layer of the system (Morris and Stormont 1999).

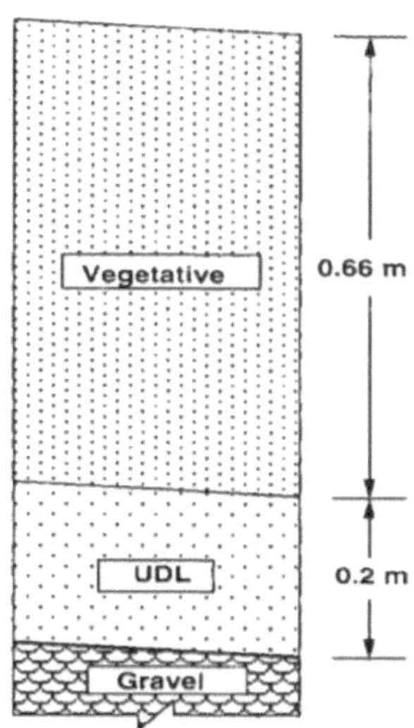

Figure 2.19 Unsaturated drainage layer (UDL) (Morris and Stormont 1999)

2.10 Governing equations for transient water flow in unsaturated soils

This section is about the transient water flow (unsteady-state) at which there is a change in moisture content with time that takes place within the unsaturated soil. Similar to the steady state flow problems, the transient water flow will be affected if there is a difference in the hydraulic conductivity in the x- and y-directions at any point in the soil profile (i.e., anisotropy). In addition, the coefficient of permeability varies with respect to space due to the change in matric suction within the soil profile (hydraulic conductivity is a function of matric suction). Because the matric suction at a point in the soil profile varies with time in a transient analysis, the hydraulic conductivity also becomes a function of time.

It is also important to take into consideration the change in water storage capacity in heterogeneous soils when considering the transient water flow.

The water flow in unsaturated soils is affected by many factors. The driving potential that allows water to flow through unsaturated soils is a function of many factors, such as water content variations, matric suction gradient, and hydraulic head gradient (Fredlund et al. 2010).

In continuum mechanics, the fundamental physics laws are the conservation of mass, the conservation of momentum, and the conservation of energy. The law of conservation of mass, applied to water flow through soils, states that for given representative elemental volume of soil "the rate of water loss or gain is equal to the net flux or the difference between the inflow and outflow" (Lu and Likos 2004b). A representative elementary volume of soil is shown in Figure 2.20.

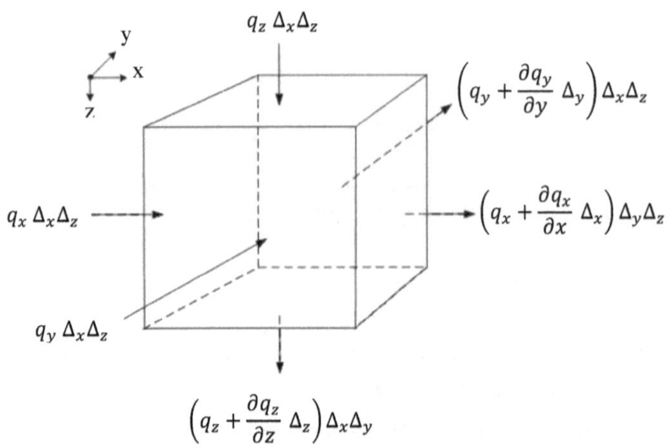

Figure 2.20 Representative elemental volume of soil and continuity requirements for fluid flow
(after Lu and Likos 2004b)

The water inflow into the soil element is expressed by the following equation.

$$q_{in} = \rho \left(q_x \Delta_y \Delta_z + q_y \Delta_x \Delta_z + q_z \Delta_x \Delta_y \right) \tag{2.30}$$

The total outflow is expressed by the following equation,

$$q_{out} = \rho \left[\left(q_x + \frac{\partial q_x}{\partial x} \, \Delta_x \right) \Delta_y \Delta_z + \left(q_y + \frac{\partial q_y}{\partial y} \, \Delta_y \right) \Delta_x \Delta_z + \left(q_z + \frac{\partial q_z}{\partial z} \, \Delta_z \right) \Delta_x \Delta_y \right]$$

$$\tag{2.31}$$

where ρ is the density of water (kg/m³); $q_x, q_y, and\ q_z$ are fluxes in x-, y-, and z-directions, respectively (m/s).

The rate of water mass that is lost or gained by the soil element through the transient water flow process is expressed by the following equation,

$$\frac{\partial(\rho\theta)}{\partial t}\ \Delta_x\Delta_y\Delta_z \tag{2.32}$$

To satisfy the law of conservation of mass, Eq. 2.32 should be equal to the net water flux as expressed in the following equation,

$$-\rho\left(\frac{\partial q_x}{\partial x}+\frac{\partial q_y}{\partial y}+\frac{\partial q_z}{\partial z}\right)=\frac{\partial(\rho\theta)}{\partial t} \tag{2.33}$$

The above equation is considered the governing equation for transient fluid flow in soil for both saturated and unsaturated conditions.

The equations describing the conservation of momentum and the conservation of energy can be found in any continuum mechanics book.

The solution of the governing equations requires that the relevant constitutive relations are known. In relation to water flow in unsaturated soils, Darcy's law is used by considering the hydraulic conductivity as a function of soil suction or suction head. Darcy's law is a constitutive relation as expressed in Eq. 2.34.

$$q_{x=-k_x}(h_m)\,\frac{\partial h}{\partial x} \qquad q_{y=-k_y}(h_m)\,\frac{\partial h}{\partial y} \qquad q_{z=-k_z}(h_m)\,\frac{\partial h}{\partial z} \tag{2.34}$$

where h_m is the matric suction head; $k(h_m)$ is the unsaturated hydraulic conductivity function. Substituting Eq. (2.34) into Eq. (2.33) and assuming a constant water density leads to,

$$\frac{\partial}{\partial x}\left[k_x(h_m)\,\frac{\partial h_w}{\partial x}\right]+\frac{\partial}{\partial y}\left[k_y(h_m)\,\frac{\partial h_w}{\partial y}\right]+\frac{\partial}{\partial z}\left[k_z(h_m)\left(\frac{\partial h_w}{\partial z}+1\right)\right]=\frac{\partial\theta}{\partial t} \tag{2.35}$$

The additional term that can be seen in the z-coordinate direction is due to the presence of the elevation head. The right side term is rewritten in terms of matric suction head as,

$$\frac{\partial \theta}{\partial t} = \frac{\partial \theta}{\partial h_m} \frac{\partial h_m}{\partial t} \tag{2.36}$$

where $\frac{\partial \theta}{\partial h_m}$ is the slope of the relationship between volumetric water content and suction head. This slope is determined directly form the soil-water characteristic curve. The specific moisture capacity (c) is referred to this slope and can be written as a function of suction or suction head as shown in the following equation.

$$c(h_{m)=} \frac{\partial \theta}{\partial h_m} \tag{2.37}$$

Substituting Eqs. (2.36) and (2.37) into Eq. (2.35) leads to the governing equation for the transient flow at the unsaturated soil.

$$\frac{\partial}{\partial x}\left[k_x(h_m) \frac{\partial h_w}{\partial x}\right] + \frac{\partial}{\partial y}\left[k_y(h_m) \frac{\partial h_w}{\partial y}\right] + \frac{\partial}{\partial z}\left[k_z(h_m) \left(\frac{\partial h_w}{\partial z} + 1\right)\right] = c(h_m) \frac{\partial h_m}{\partial t} \tag{2.38}$$

2.11 Transient heat flow in unsaturated soils

2.11.1 Two dimensional heat flow

The heat transfer differential equation (Eq. 2.39) in two dimensions for unsaturated soils can be found in Fredlund et al. (2012c).

The net heat flux through a representative elemental volume is equal to the volumetric heat capacity of the soil multiplied by the change in temperature with respect to time in order to satisfy the law of conservation of the energy. The term on the right side of the equation is called a "heat storage" term.

$$\lambda \frac{\partial^2 T}{\partial x^2} + \frac{\partial \lambda}{\partial x} \frac{\partial T}{\partial x} + \lambda \frac{\partial^2 T}{\partial y^2} + \frac{\partial \lambda}{\partial y} \frac{\partial T}{\partial y} = \zeta \frac{\partial T}{\partial t} \tag{2.39}$$

where λ is thermal conductivity of the soil; ζ is the volumetric heat capacity, and T is the temperature.

2.12 Coupled heat transfer and moisture flow in unsaturated soil

The transport of liquid water, water vapor and heat in the soil have been studied by many researchers over the past several decades (Bach 1992; Hussain 1997; Bittelli et al. 2008; Adala et al. 2009; Sakai et al. 2009; Bahador et al. 2014). The intent of these studies has been to better understand the effects of fluid flow and heat transfer in the behaviour of unsaturated soils.

Philip and Vries (1957) described the moisture and heat transfer in porous media under moisture and temperature gradient considering the interaction between vapor, liquid, and solid phases.

The behavior of an unsaturated soil column subjected to variations in water vapor, liquid movement, and heat transfer was studied by Bach (1992), Hussain (1997), Bittelli et al. (2008), Adala et al. (2009), Sakai et al. (2000), and Bahador et al. (2014). They reached the conclusion that the heat transfer through the system caused a vapor movement from the high-temperature zone (at the bottom of the box) toward the lower temperature zone at the top of the box.

Sakai et al. (2009) studied the movement of water, vapor, and heat in a sandy column test. They concluded that the water vapor migrated from the hot boundary to the cold boundary end and then condensed.

Bahador et al. (2014) studied the effect of the temperature on the suction values and the moisture distribution (i.e., movement of water in the soil column) in case of using geotextile as a capillary barrier layer. They reached the conclusion that the heat transfer through the system caused a moisture movement from high temperature zone (at the bottom of the soil column) toward the lower temperature zone at the top of the soil chamber.

Bach (1992) studied the effect of temperature gradient on the movement of the soil moisture He indicated that movement of water from warmer to cooler regions had a significant effect on the overall soil-water movement in unsaturated soils. The moisture moved from the hot region to the cold region through the vapor phase in the closed column (i.e., no water inflow or outflow) when the soil column was subjected to temperature gradient. At the cold end of the soil column, condensation took place.

More recent research works by Hussain (1997) and Adala et al. (2009) indicated that liquid transport is due to combined gravity effect and transport associated with matric suction gradients caused by moisture content and temperature gradients. King (1892) argued that he was right (as cited in Roshani, 2014) that the variation of temperature could have an effect on the water holding capacity of the unsaturated soil (matric suction).

Bittelli et al. (2008) indicated that the vapour flow plays a role in the soil mass and energy transfer (i.e., phase change from a gas to a liquid). They indicated that the degree of soil saturation can have a significant effect on the air filled porosity and vapor diffusion (it is equal to zero in a fully saturated soil). The vapour flux increases with increasing air porosity, in other words, when the soil is getting dry. Equation 2.40 for the soil vapour flux describes the last mentioned mechanism.

$$D_v = D_0\varepsilon(x_a) \tag{2.40}$$

where D_v is soil vapor diffusivity (m²/sec), D_0 is the binary diffusion coefficient for a gas (m²/sec), $\varepsilon(x_a)$ is a soil parameter that depends on the air filled porosity (x_a), and the soil geometrical properties. Air filled porosity (x_a) is the difference between the volumetric water content at saturation θ_s and the actual volumetric water content (θ).

The parameter $\varepsilon(x_a)$ is calculated using the following equation,

$$\varepsilon(x_a) = \beta x_a^m \tag{2.41}$$

where β and m are empirical parameters that depend on the shape of the soil particles.

The two most important factors that affect heat transfer are the thermal conductivity of the soil and the volumetric heat capacity of the soil. Additionally, the thermal conductivity in an unsaturated soil is a function of SWCC as shown in Figure 2.21.

The degree of saturation is a key parameter to estimate the thermal conductivity of the soil. In contrast, the degree of saturation is also related to the matric suction. Therefore, the thermal conductivity can be influenced by matric suction as shown in Figure 2.21.

The relationship between the thermal conductivity and the matric suction becomes clearer when the coupled thermal and moisture flow analysis is considered as can be expressed in Eq. 2.42. Accordingly, the SWCC can be used as a tool not only to calculate the moisture flow properties but, also the heat flow properties such the thermal conductivity.

The thermal conductivity, λ of a soil consisting of soil particles, water, and air can be expressed as follows (de Vries, 1963):

$$\lambda = \frac{f_p\,\theta_p\,\lambda_p + f_w\theta\lambda_w + f_a\theta_a\lambda_a}{f_p\theta_p + f_w\theta + f_a\theta_a} \tag{2.42}$$

where:

λ is the thermal conductivity of the unsaturated soil

f_p, f_w, f_a are weighting factors for the solid particles, water, and air phases, respectively,

$\theta_p, \theta, \theta_a$ are percentage of total volume that is solid particles, water, and air, respectively, and

$\lambda p, \lambda w, \lambda a$ are thermal conductivities of the solid particles, water, and air phases, respectively

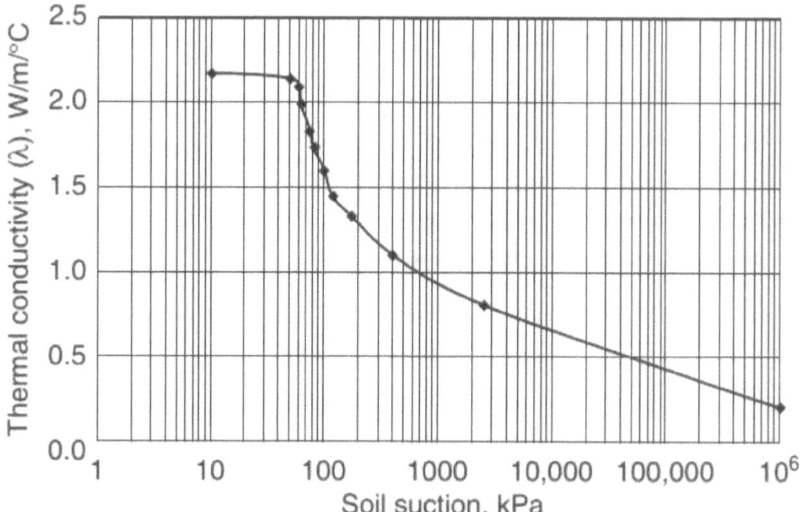

Figure 2.21 Thermal conductivity versus matric suction for silica flour (Fredlund et al. 2012 b)

In Figure 2.21, it can be seen that matric suction decreases as the thermal conductivity increases. As the matric suction of the fine-coarse interlayer should be kept higher than the WEV to prevent percolation, the corresponding thermal conductivity should be in the range of 0 to 0.75 W/m/°C. To achieve this, the temperature of the soil should also be taken into account as it has a significant effect on the thermal conductivity. It can be concluded the air temperature should be within an acceptable range to provide efficient matric suction (higher than WEV) in terms of thermal conductivity.

Gurr et al. (1952) studied the movement of water in the sealed soil column under the effect of temperature gradient. They indicated that the water vapor moves from high temperature zone to the cold zone within the soil column and concluded that the water movement in the soil column is affected by many factors namely, the initial water content, the bulk density, the temperature gradient, and the mean temperature.

Their explanation for this movement was that as water is evaporating at the hot zone, it moves in the vapor phase toward the cold zone under effect of the temperature gradient in the soil column.

Taylor and Cavazza (1954) studied the movement of water in a closed soil column under the effect of thermal gradient. They indicated that the movement of water from the hot region to the cold region was in vapor phase.

Bittelli et al. (2008) indicated the movement of heat and moisture in the soil structure are coupled, that is, the variation of temperature (the thermal gradient) in the soil affects the moisture distribution in the soil profile. Correspondingly, the moisture (in vapor phase) moves from the hot boundary toward the cold boundary.

When multiphase flow in the soil mass is observed and there is a vapor flow combined with thermal gradient, it is necessary to consider the coupled transport concept.

The partial differential equation for transient vapour flow was proposed (Dakshanamurthy and Fredlund 1981; Fredlund and Dakshanamurthy 1982; Wilson et al, 1994; Fredlund et al .1999; Bittelli et al. 2008) to describe the vapour flow within the unsaturated soil using Fick's law.

$$qv = -Dv \left(\frac{dPv}{dy}\right) \qquad (2.43)$$

$$Pv = Pvs \; x \; hr \qquad (2.44)$$

$$hr = \exp\left(\frac{Wv * \Psi}{R * T}\right) \qquad (2.45)$$

where qv is water vapour flux (kg/ (m² s); Pv is partial pressure due to water vapour (kPa) within the voids in the unsaturated soil; Pvs is the saturated vapor pressure (kPa) of the soil water at soil temperature (obtained from Kaye and Laby 1911); hr is the relative humidity; Ψ is the total potential in the liquid water phase expressed as equivalent matric potential (m). Wv is the molecular weight of water = (0.018 kg/mol); R is the universal gas constant = 8.314 J/ (mol K).

$$Dv = (\alpha)(\beta)\left[Dvap\left(\frac{Wv}{RT}\right)\right] \qquad (2.46)$$

$$\alpha = \beta^{2/3} \qquad (2.47)$$

$$\beta = (1\text{-}S)*n \qquad (2.48)$$

$$Dvap = 0.229 * 10^{-4}\,[1 + (T/273)]^{1.75} \qquad (2.49)$$

where Dv is the diffusion coefficient of the water vapour through soil (kg.m / (kN.s)); α is tortuosity factor of soil; β is cross sectional area of soil available for vapour flow; S is the degree of saturation; n is porosity (0.41); $Dvap$ is molecular diffusivity of water vapour in air (m²/s).

Fredlund et al. (2012) indicated that the amount of vapor in the pore air phase depends on the total suction and soil temperature (the thermodynamic relationship). Accordingly, any change that occurs in the total suction or the temperature causes mass transfer that takes place between the gaseous and liquid phases.

By neglecting the effect of the osmotic suction, and assuming the air pressure is equal to the atmospheric pressure, the coupled moisture and heat flow in the soil can be expressed using Kelvin's equation.

$$\psi = \frac{RT\rho_w}{\omega_v} \ln(\frac{u_v}{u_{v0}})$$
(2.50)

where Ψ is the soil suction or total suction (kPa); R is the universal gas constant [8.31432 J/(mol K)]; T is absolute temperature (K); ρ_w is density of water (998 kg/m³); ω_v is molecular mass of water vapor (18.016 kg/kmol); u_v is partial pressure of pore-water vapor (kPa); and u_{v0} is saturation pressure of water vapor over a flat surface of pure water at the same temperature (kPa). The term $\left(\frac{u_v}{u_{v0}}\right)$ is equal to the relative humidity.

2.13 Knowledge gap

Based on the literature review, it was concluded that there are only a limited number of studies examining the effect of air flow on the performance of capillary barrier systems. In addition, there is no published record of any investigation aimed at evaluating the effect of heated air flow on the performance of capillary barrier systems. The author concludes that there is a need for both experimental and numerical studies to fill the identified knowledge gap.

To achieve this objective, an experimental setup was built to study the effect of the heated airflow on different parameters (i.e. Volumetric water content and matric suction), related to the behaviour of the capillary systems. The study was extended to include a modeling phase to validate the approach, whereas no numerical analysis was done in the past to study the effect of using heated airflow on the performance of the capillary barrier system. To investigate the applicability of the developed numerical model, it was applied to two case studies described in the literature, and the results obtained clearly demonstrated that the performance of a capillary system can be improved using the proposed approach.

One pipe was utilized to introduce airflow in the current study, while two pipes were used in the past (a pressure line and a vacuum line) for this purpose. Previous studies employed active systems to force air to circulate through the capillary barriers, but this can be eliminated with the current approach using heated airflow generated by a passive system (Ex: solar thermal energy). Therefore, the author believes that the current approach is simpler to implement in the field and can offer some cost savings in construction as well as in operation.

In the past, the water storage capacity of capillary barriers was improved by either increasing the thickness of the fine-grained soil layer, constructing more than two layers, or by extending the length of the slope to increase the diversion length. The proposed approach achieves this objective by allowing heated air to circulate through the system.

The novelty of the current approach is demonstrated in Table 2.2, which provides a comparison between other drainage techniques and the self-dewatering technique (topic of the present research using heated air flow).

Table 2.2 Comparison between self-dewatering and other drainage techniques

Technique	Factors affecting operation	Principles of operation	Performance
Slope capillary barrier	Slope angle, layers thickness, soil materials	Geometry	Short -term drying (during the precipitation event)
Unsaturated drainage layer (UDL)	The materials of the layer, the slope of the fine-coarse interface, unsaturated flow characteristics	Geometry and structure	Short -term drying (during the precipitation)
Dry barrier system	Wind speed, RH of the air, pipe spacing, pressure gradient	Blower, fan or wind-powered chimney. Bulk air flow. Two pipelines are needed.	Short term drying (seasonal technique)
Self-dewatering technique	Air flow rate, air temperature,	Currently: circulation of preheated air In the future: natural processes (green energy, e.g., solar radiation).	Long-term drying (during and after precipitation)

2.14 Summary

In this chapter, all theoretical background and equations relevant to the present study have been summarized. A large portion of the chapter is devoted to the material covered in the section of the capillary barriers in the textbook by Lu and Likos (2004). The experimental work by Wilson (Wilson et al. 1995; Wilson et al. 1997) provides justification for using high suction values at very low relative humidity.

The preceeding literature review described various methodologies used to keep a capillaary barrier to continue functioning effectively. In these methods, the circulation of air through the coarse grained soil layers of capillary barriers was used. There was no method described in the literature that woud use heated air flow in the coarse grained soil layer to improve the effectiveness of capillary barriers.

As described in Chapter 1, the main objective of the present research is to find out whether heated air flow will improve the performance of capillary barriers. The heated air flow is a self-dewatering method and, therefore, it can be considered as a long-term solution to keep the soil dry during and after a precipitation event.

CHAPTER 3

MODEL SCALE LABORATORY EXPERIMENTS: HEATED AIR FLOW THROUGH COARSE GRAINED SOIL LAYER OF A CAPILLARY BARRIER SYSTEM

3.1 Introduction

Several types of tests were conducted to determine the geotechnical properties of a sandy soil which was used in all experiments described in this chapter. It is assumed that this is the soil type to be used in the construction of the coarse grained soil layer of a capillary barrier system. In the determination of the soil properties, two things were taken into consideration: (1) the soil is unsaturated and at various degrees of saturation, and (2) the temperature is an important factor influencing the soil properties.

The model scale labrotory experiments presented in this chapter include: (a) heated air flow tests and, (b) non heated air flow tests. The purpose of these experiments was to determine the effect of air heating on the performance of the coarse grained soil layer of a capillary barrier. The material used to construct the coarse grained soil layer was the Silica Sand G-20 described in Section 3.2.1. If a two layer capillary barrier were to be constructed, a fine grained soil layer consisting of silt or clay would be placed on top of the sand layer.

In the following, the experimental set-up and sensors used to measure relevant physical quantities are described. The test results and a discussion are presented subsequently in Section 3.6.

All tests were performed at the University of Ottawa Geotechnical Laboratories.

3.2 Geotechnical properties of a silica sand laboratory experiments for the measurement of geotechnical properties of a silica sand

3.2.1 Description of the soil

The soil utilized in this study is called Silica Sand-G20. It is sold by Beel and Meckenzie Co. Ltd. Figure 3.1 shows a picture of the sand.

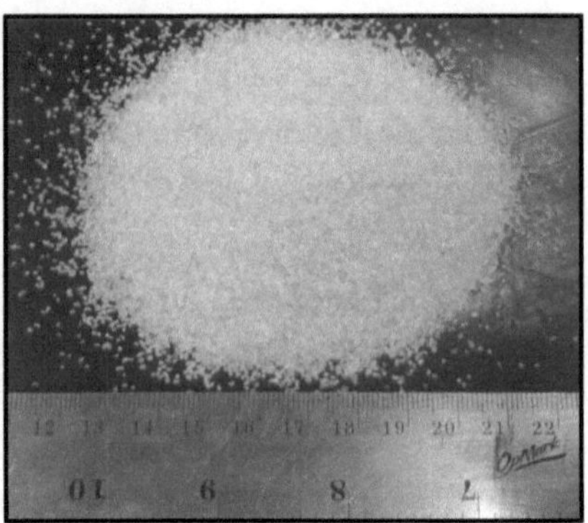

Figure 3.1 Silica Sand-G20

3.2.2 Properties of Silica Sand-G20

Table 3.1 summarizes the chemical composition of the silica sand used in the experimental program. The water pycnometer method was followed to measure the specific gravity of the soil. The minimum dry density and the unit weight of the soil were determined using a funnel to place material in a mold (Method A- D4254-16). Standard vibratory table method was used to measure the maximum density and unit weight. The Constant Head test procedure (ASTM D 2434-68) was followed to measure the hydraulic conductivity of the saturated soil as shown in Figure 3.2. Some index and mechanical properties of the silica sand, as published in ASTM standards, are presented in Table 3.2.

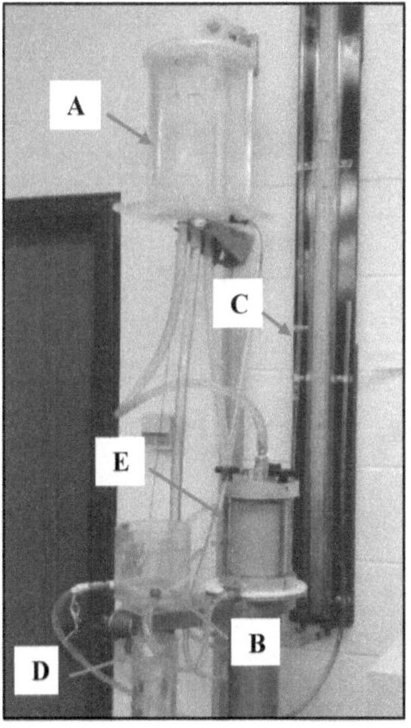

Figure 3.2 The hydraulic conductivity test set up (A) Head tank, (B) Overflow funnel, (C) Piezometers, (D) Collector tube, (E) Permeameter cell

The hydraulic conductivity determined in this test was for the saturated sand. For the unsaturated conditions, the hydraulic conductivity was determined from the equation of van Genuchten (1980).

$$K_W = K_S \frac{\left[1 - \left(a\Psi^{(n-1)}\right)\left(1 + (a\Psi^n)^{-m}\right)\right]^2}{\left[\left((1 + a\Psi)^n\right)^{\frac{m}{2}}\right]} \tag{3.1}$$

Where k_s is saturated hydraulic conductivity, a, n, m are curve fitting parameters, n is $1/(1\text{-}m)$, and Ψ is required suction range.

Table 3.1 Chemical composition of Silica Sand-G20 (Beel and Meckenzie Co. Ltd)

Component	Formula	%
Silicon dioxide	SiO2	99.6
Iron Oxide	Fe2O3	0.02
Aluminum oxide	Al2O3	0.20
Calcium oxide	CaO	0.01
Magnesium oxide	MgO	0.01
Potassium oxide	K2O	0.01
Sodium oxide	Na2O	0.01
Loss in Ignition	LOI	0.1
Colour		White to off-white
Grain shape		Subangular

Table 3.2 Some index and mechanical properties of silica sand

Test	Value	Unit	ASTM Standard
Specific Gravity (GS)	2.55	-	D854-98
Minimum unit weight (γ_d min)	13.6	kN/m³	D4254-16
Maximum unit weight (γ_d max)	15.740	kN/m³	D4253-16
Minimum dry density (ρ_d min)	1.388	g/cm³	D4254-16
Maximum dry density (ρ_d max)	1.605	g/cm³	D4253-16
Hydraulic conductivity (K_s) (Soil is saturated)	0.068	cm/sec	D 2434-68
Void ratio (e_0) max	0.84	-	D4254-16
Void ratio (e_0) min	0.59	-	D4253-16

3.2.3 Particle-size distribution curve (PSD)

For the sieve analysis, the standard test method, ASTM D6913-04 (ASTM 2004), was followed to obtain the particle-size distribution curve, as shown in Figure 3.3. The Unified Soil Classification System (USCS), as summarized in ASTM D 2487-98 (ASTM 1998), was applied to classify the material used in the experimental program, as shown in Table 3.3.

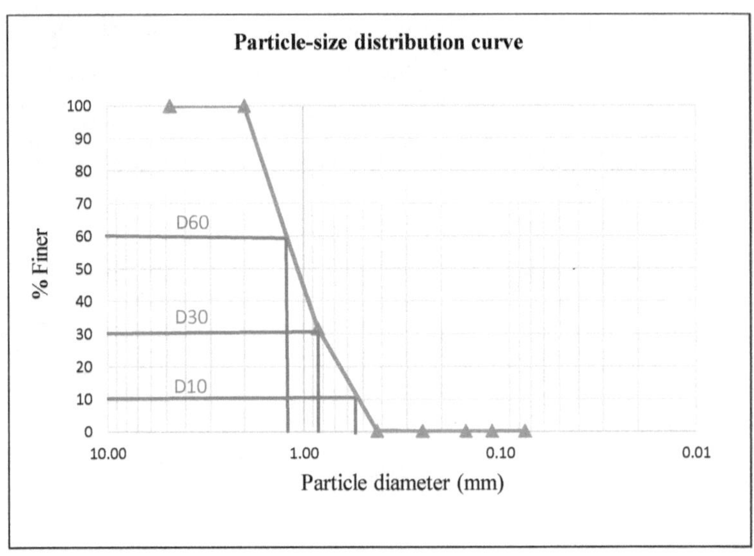

Figure 3.3 Particle-size distribution curve for the silica sand

Table 3.3 USCS for Silica Sand

Silica sand	
ASTM D2487 – 98	
Coefficient of uniformity, Cu	0.42
Coefficient of curvature Cc	0.99
USCS	SP (Poorly graded sand)

Based on the Unified Soil Classification System (USCS), the soil used in this research was classified as "medium, poorly graded sand".

3.2.4 Soil-Water Characteristic Curve (SWCC)

A hanging column test was performed to obtain the Soil-Water Characteristic Curve (SWCC) for the silica sand, used in the experiments. The hanging column set-up allows application of low vacuum pressures in controlled small increments and is suitable for measuring matric suction in the range 0-80 kPa. ASTM D6836-02 (ASTM 2008) procedure were followed in preparing the experimental test set-up. The hanging column apparatus was combined with a Tempe cell, which is also used for measuring low range matric suction. The Tempe cell is a metal cylinder that contains a high air entry ceramic disk (100 kPa). The initially saturated soil sample was placed on top of the saturated high air entry ceramic disk, and the Tempe cell was connected to the hanging column apparatus to apply vacuum pressure. The first pressure increment was selected to be equal to the desired lowest matric suction value. As soon as the pressure was applied, the water started draining from the sample through the ceramic disk and the process continued until equilibrium condition was reached. At equilibrium, the matric suction of the soil was equal to the applied vacuum pressure. The volume of drained water from the sample was measured and the Volumetric Water Content (VWC) corresponding to the measured matric suction was calculated. Test was repeated at increasing vacuum pressures and the measured matric suctions and VWCs were plotted to obtain the SWCC for the sand. Figure 3.4 shows the test set-up and Figure 3.5 presents the Soil-Water Characteristic Curve (SWCC) for the silica sand. The measured data points and the fitting curve parameters for Fredlund and Xing (1994) are summarized in Table 3.4.

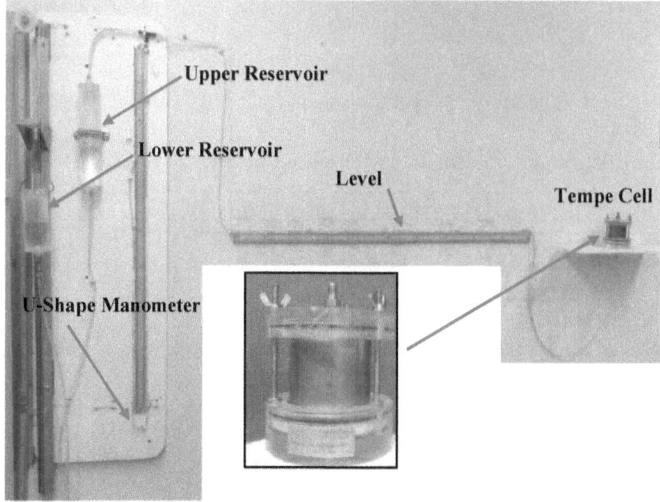

Figure 3.4 Hanging column test setup

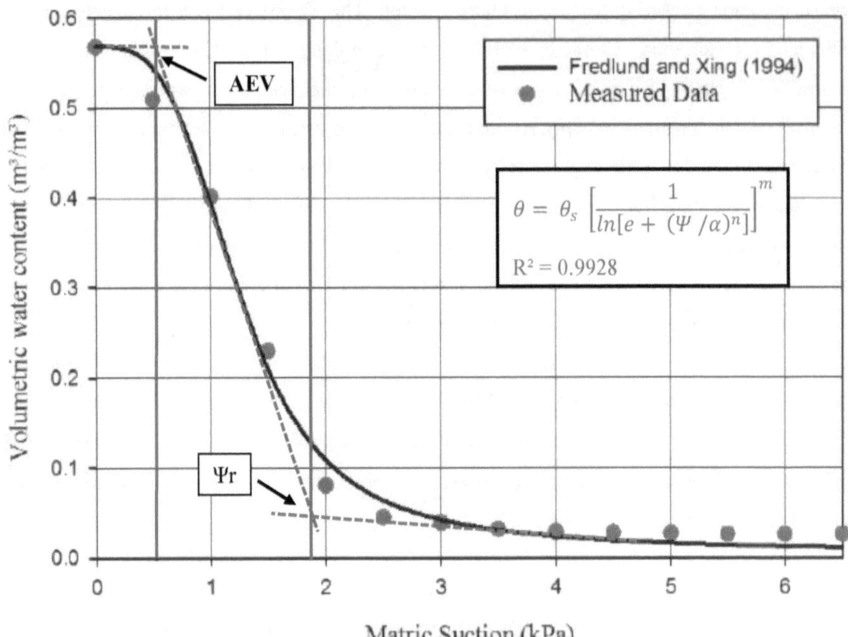

Figure 3.5 The soil-water characteristic curve (SWCC) of Silica Sand G-20

Table 3.4 Parameters of soil water characteristic curve (SWCC) of silica sand

SWCC (Drying curve)							
Fitting parameters				θs	AEV (Ψa)	θr	Ψr
α (kPa)	n	m	R^2				
1.2212	3.3321	2.2787	0.9928	m³/m³	kPa	m³/m³	kPa
				0.5680	0.7	0.05	1.9

R^2: coefficient of determination of the curve

3.3 Experimental set-up

The physical model consists of two main components: the air heating box and the sand box, as shown in Figure 3.6.

The air heating box (AHB) is a steel box containing a ceramic heater that was used to provide a source of heated air. The dimensions of the air heating box are: 69.4 cm length, 36.2 cm height, and 45.7 cm width. The heating box was insulated with 4.9 cm thick insulation (Polystyrene rigid insulation). The ceramic heater, with a power output of 1500 Watt, was attached to the box to heat the air inside the box. A centrifugal fan was used to give directionality to the air flow (i.e., to direct and concentrate the heated air flow into a PVC pipe, which carried the heated air to the sand box.). A small, square window was cut to allow external air to enter the heating box (Figure 3.6).

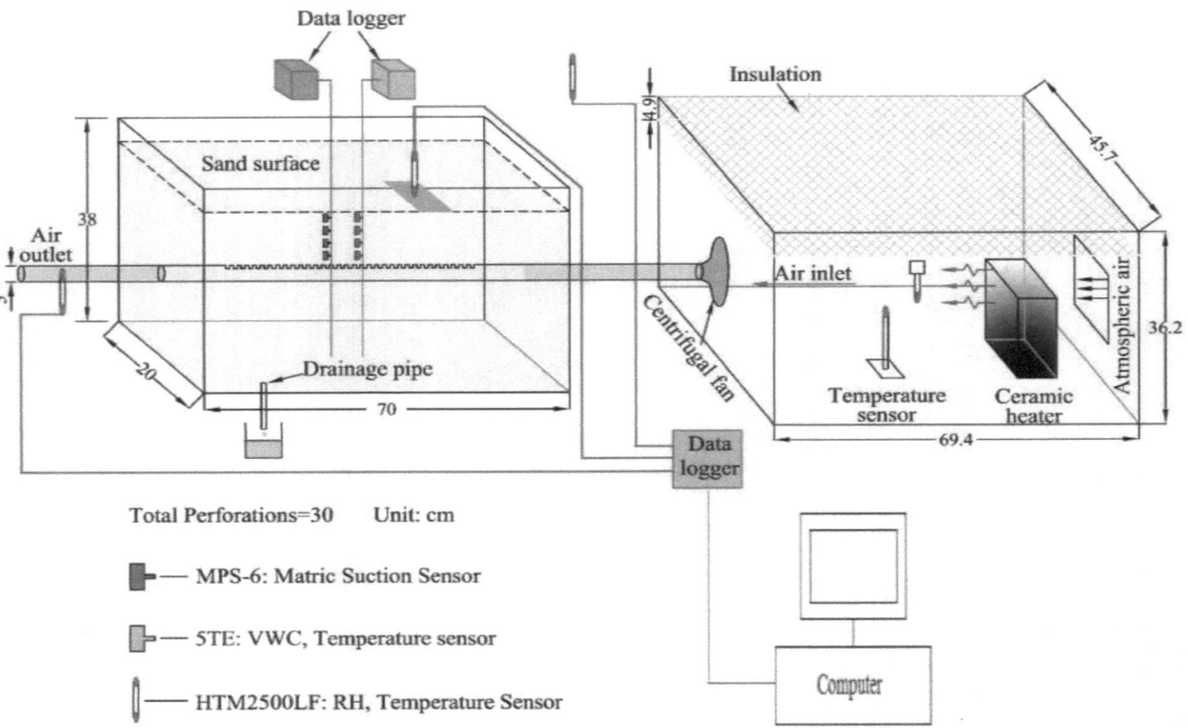

Figure 3.6 Schematic view of the physical model

The sand box was built using 5.6 mm-thick acrylic sheet with the following dimensions: 38 cm height, 70 cm length, and 20 cm width. The PVC pipe connecting the AHB to the sand box was 1 m long and 5 cm in inner diameter (6 mm outer diameter). Thirty perforations were drilled along the upper side of the pipe, as shown in Figure 3.7. The diameter of the perforations was 7.5 mm. In earlier experiments, the diameter of the perforations was 3.86 mm. It was realized that increasing the perforation diameter from 3.86 mm to 7.5 mm accelerated vapour flow from the soil to the pipe and, therefore reduced the time spent for the equilibrium states to be reached. Air permeable fabric (chiffon fabric) was used as a filter to prevent sand from migrating into the pipe. At a later stage of the experimental work, several other modifications were made for the locations of instruments/sensors, as will be explained in Section 3.4.1.

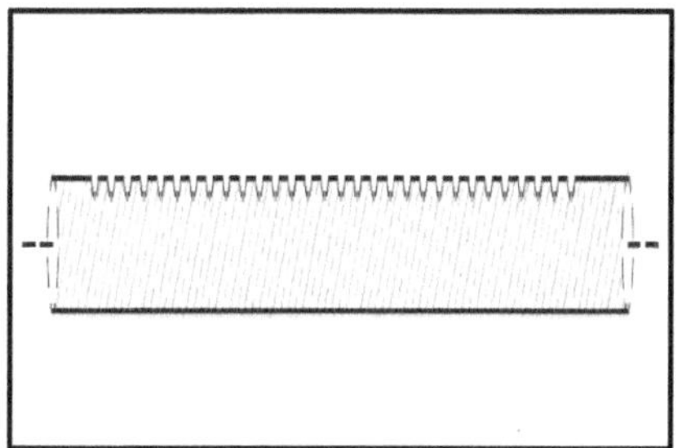
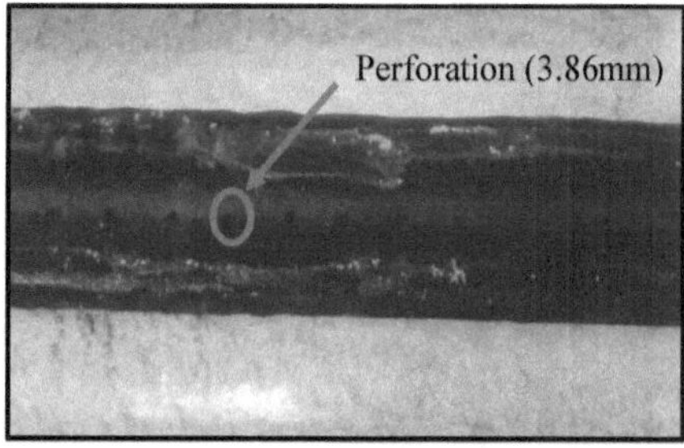

Figure 3.7 Schematics of perforation and a picture of the perforated pipe

3.4 Sensors used in the experiments

Sensors were used to measure the effect of heating on the temperature distribution, volumetric water content, and suction in the sand mass. Four commercially available sensors manufactured by Decagon Devices Inc. were used in these tests. These sensors were: 5TE (frequency domain reflectometry + thermistor + electrical conductivity sensor) used to measure volumetric water content, temperature sensors, relative humidity sensors and MPS-6 sensors to measure suction. The MPS-6 sensor measures suction indirectly by monitoring the volumetric water content of a ceramic matrix enclosing a frequency domain reflectometry sensor similar to that used in the 5TE. Figure 3.8 shows a picture of all the instruments. Manufacturers recommended range of measurements that can be taken by the sensors are listed in Table 3.5.

Figure 3.8 Various types of sensors used in the heated air flow tests: (A) MPS-6 matric suction sensors, (B) 5TE volumetric water content sensors, (C) Relative humidity sensor located at the pipe outlet, (D) Relative humidity sensor located at soil surface, (E) Relative humidty sensor located at AHB, and (F) Temperature sensor located at AHB

Table 3.5 The specification of the instruments used in the study

Model	Range		Accuracy		Sensor
MPS-6	Matric Suction (kPa)	Temperature (°C)	Matric Suction (kPa)	Temperature (°C)	
	9-100,000	-40 to 60	±(10% of reading + 2 kPa) from 9 to −100 kPa	±1	
5TE	VWC (m^3/m^3)	Temperature (°C)	VWC (m^3/m^3)	Temperature (°C)	
	Apparent dielectric permittivity (εa): 1 (air) to 80 (water)	-40 to 60	± 0.03 (± 3% VWC) typical	± 1	
Anemometer (HHF801)	Temperature (°C)	Air velocity (m/s)	Temperature (°C)	Air velocity (m/s)	
	0 to 80	0.8 to 12.0	±0.8	±(2% + 0.2)	
HTM2500LF	Relative Humidity (%)	Temperature (°C)	Relative Humidity (%)		
	-40 to 85	0 to 100	±3% to ±5%		

3.4.1 Locations of the sensors placed in the sand box

Originally, when the experimental work started in the lab several years back, the sensors were placed in the sand mass along vertical lines at 4 cm, 8 cm, 12 cm, and 16 cm above the top of the buried pipe. This arrangement of sensors is shown in Figure 3.6. In order to make the sensor locations clearer, an enlarged view of the sand box (before the placement of sand) is shown in Figure 3.9.

Figure 3.9 Enlarged view of the sand box

In Figure 3.6, four small red squares placed above each other 4 cm apart indicate the locations of the suction (plus temperature) sensors. Similarly, four small blue squares placed above each other 4 cm apart indicate the locations of the volumetric water content (plus temperature) sensors. In further references to this sensor arrangement in this book, this original arrangement for sensor locations is called "ASL-type #1".

Although this arrangement of sensors seemed suitable, there were concerns that it may cause problems, because downward drainage of water in the soil mass was interrupted, vapour flow was impeded, and temperature distribution was also altered.

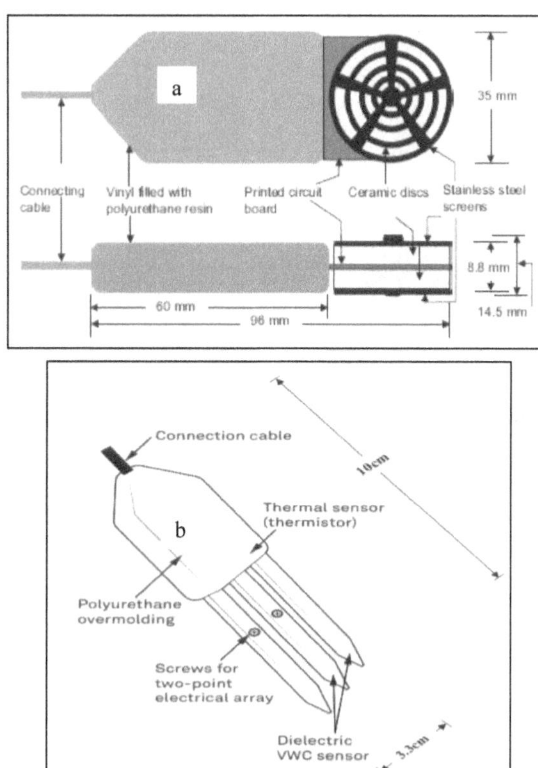

Figure 3.10 Dimensions of sensors (a) MPS-6: Sensor to measure matric suction (Tripathy et al. 2016), (b) 5TE: Sensor to measure volumetric water content (modified after Meter group, Inc., 2012)

Figure 3.10 shows the dimensions of these sensors. Because of their large size, measurements taken for the suction and the volumetric water content do not represent values at a single point, but, rather at two different points about 3.5 cm apart. Other arrangements for sensor locations are shown in Figure 3.11 and Figure 3.12.

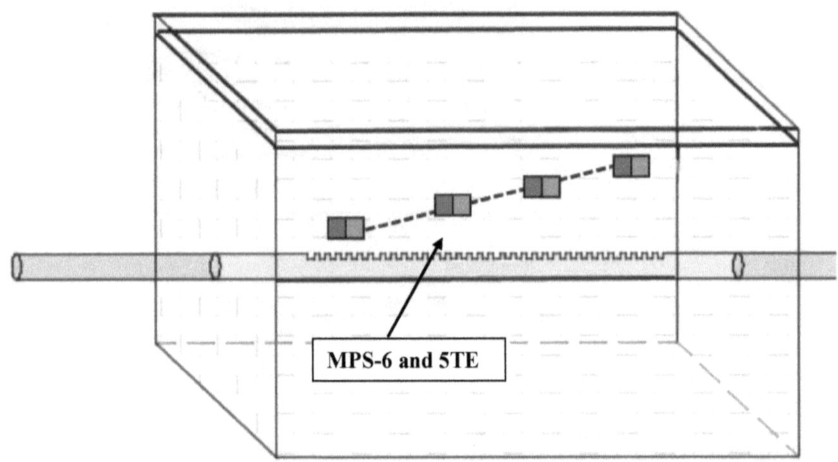

Figure 3.11 Arrangement for sensor locations (ASL-type # 2). Sensors are located on a diagonal line as shown in the figure

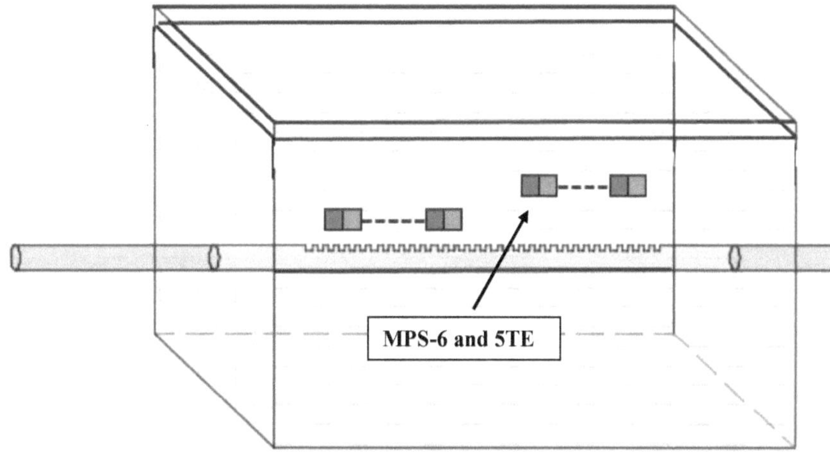

Figure 3.12 Arrangement for sensor locations (ASL-type # 3). Two pairs of sensors are located at the same elevation

3.5 Test procedures

At the beginning of each test, fully saturated sand was placed in the sand box to ensure uniform water content distribution throughout the sample. Initially, there was no heating or air flow in the pipe. Using a drainage outlet at the bottom of the sand box, water was allowed to evacuate the saturated soil mass. The drained water was collected at the bottom of the box using a collector bucket. This proses continued until no water was coming out of the drainage outlet. Therefore, at the end of this waiting period, the sand mass became unsaturated. Next, the top of the sand box was insulated by using a Polystyrene Insulation foam by Owens Corning Llc and sealed by using a plastic sheet to prevent contact between atmospheric air and the sand surface. This is the starting state for all tests described below.

3.6 Tests for the measurement of volumetric water content, matric suction, and temperature

The volumetric water content, matric suction, and temperature changes were measured in four different tests as explained below.

3.6.1 Test Number # 1: No-Air flow test

Figure 3.13 shows the test setup. Air circulation in the pipe was not allowed. The test was carried out at room temperature. Arrangement for sensor locations (ASL-type # 1) for this test is shown in Figure 3.9. The relative humidity and the room temperature were measured by the sensor (HTM2500LF) which is a product of TE Connectivity Corporation.

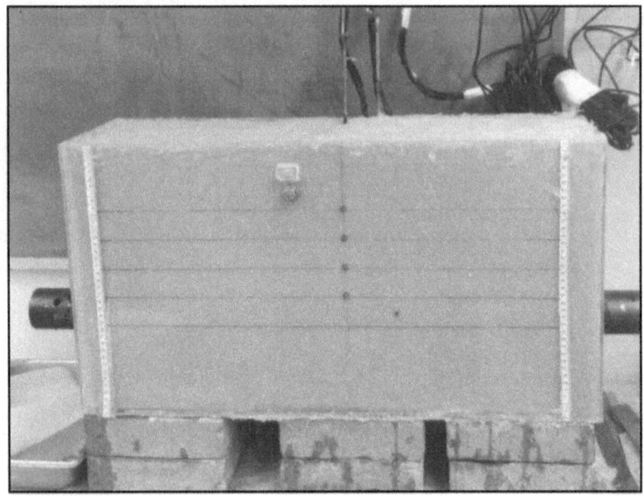

Figure 3.13 Test Number 1: Test setup for No-Air flow experiment

3.6.1.1 Measured Volumetric Water Content Values (No-Air flow test)

Figure 3.14 shows the variation in the volumetric water content with time recorded by sensors (5TE) at various distances from the pipe during the no air flow test. The duration of the test was 12 days, and readings were taken continuously. The sensors started reading VWC approximately one hour after placing the sand in the box. This time was required to finish the preparation of the test set up (i.e., to seal the sand box).

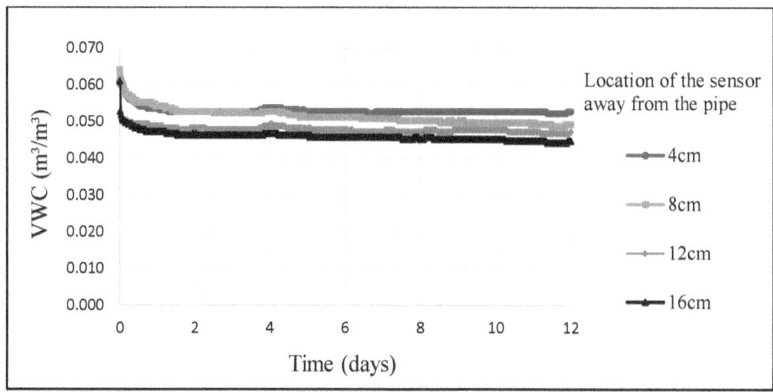

Figure 3.14 Variation of the VWC with time in No-Air flow test

The results obtained in the No-Air flow test indicate that the volumetric water content (VWC) changed only slightly and most of the changes were during the first day of the 12day-long experiment. The water content changed from 0.064 m³/m³ to 0.054 m³/m³ at 4 cm and 8 cm, respectively, and from 0.061 m³/m³ to 0.049 m³/m³ at 12 cm and 16 cm, respectively.

3.6.1.2 Measured Matric Suction Values (No-Air flow test)

Measured matric suction values are plotted in Figure 3.15. Measurements were taken continuously and recorded by a data acquisition system. It can be noted that matric suction did not show significant change, which is an indication that the VWC of the sand did not change much during the test period (12 days).

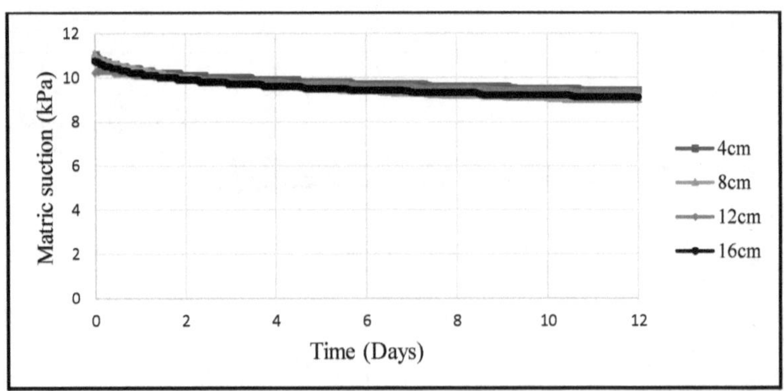

Figure 3.15 Measured matric suction values at different depths in the sand mass during the No-Air flow test

3.6.1.3 Measured Temperatures During the No-Air flow Test

Measured temperature values during this test are shown in Figure 3.16.

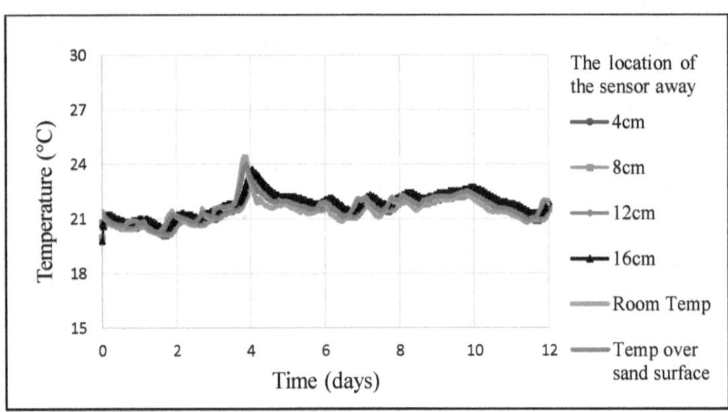

Figure 3.16 Variation of temperature vs. time at various depths during the no-air flow test

Measured temperatures at different depths in the sand mass followed the changes taking place in the room temperature.

3.6.2 Test Number #2: Heated-Air flow Test

The objective of the Heated-Air flow Test described in this section (and the sections to follow) was to determine the effect of heated-air flow on the performance of a capillary barrier system. The experimental setup is shown in Figure 3.17. Arrangements for sensor locations (ASL-type # 1) for this test is shown Figure 3.9.

Figure 3.17 Test Number #2: Test setup for "Heated-Air flow" experiments

The sand mass in the sand box represents the coarse grained soil layer of a capillary barrier. The fine grained soil layer of a capillary barrier was not included in the lab experiments. However, in the numerical simulations of lab experiments of other investigators and field tests presented in later chapters, both fine and coarse grained soil layers are present in the capillary barriers. Inclusion of a fine grained soil layer in the present experimental work would require more elaborate test set-up. The difficulties involved with sample preparation, requirements for additional instrumentation, and the choice of environmental variables would be challenging. Instead, the measurement of changes in volumetric water content and matric suction resulting from heated air flow was considered sufficient to achieve one of the objectives of the present research.

The test was conducted by using heated air flow through the perforated pipe embedded in the sand mass. The temperature of the air going through the pipe was 60°C. Arrangement for sensor locations (ASL-type # 1) for this test is as shown in Figure 3.9.

3.6.2.1 Measured Volumetric Water Content Values (Heated-Air flow test)

Figure 3.18 shows the volumetric water content vs. time at four different locations.

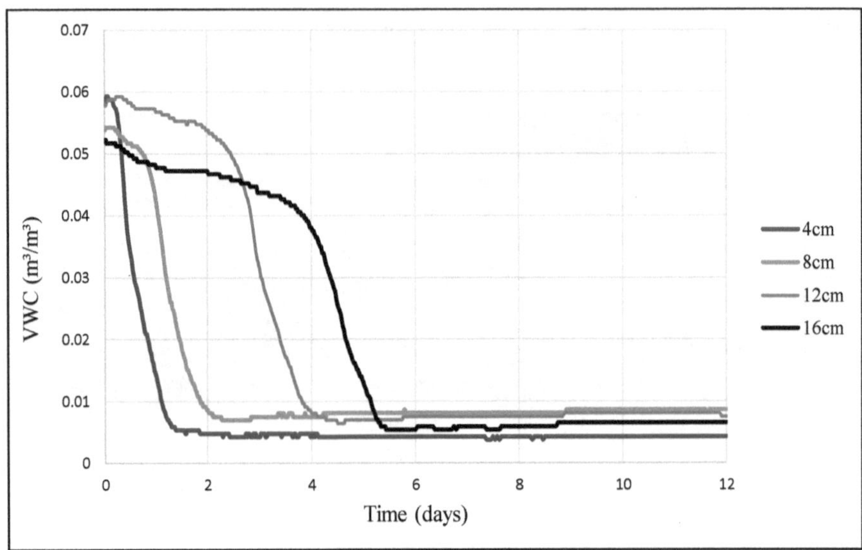

Figure 3.18 Variation of the VWC at four locations during the heated air flow test (Heated Air temperature was constant at 60°C)

It can be seen that a sharp reduction occurred in the volumetric water content values at all locations during the test. The time required for the VWC at a sensor location to reach its minimum value is related to the distance between the sensor and the top of the pipe. For example, at a point 16 cm away from the pipe, the time required for the VWC to reach its minimum value was about 5.5 days. The main observation to be made from this test was that the VWC at all points decreased significantly as a result of heated air flow through the pipe.

3.6.2.2 Measured Matric Suction Values (Heated-Air flow test)

Arrangement for sensor locations (ASL-type # 1) for this test is as shown in Figure 3.9. Measured matric suction values are shown in Figure 3.19. The corresponding VWC values are given in Figure 3.18. The increase in the matric suction begins at the sensor 4 cm above the pipe because the reduction in the VWC takes place first at this sensor location. The maximum values at the end of each line in Figure 3.19 indicate that the possible range of measurements by the sensors is reached.

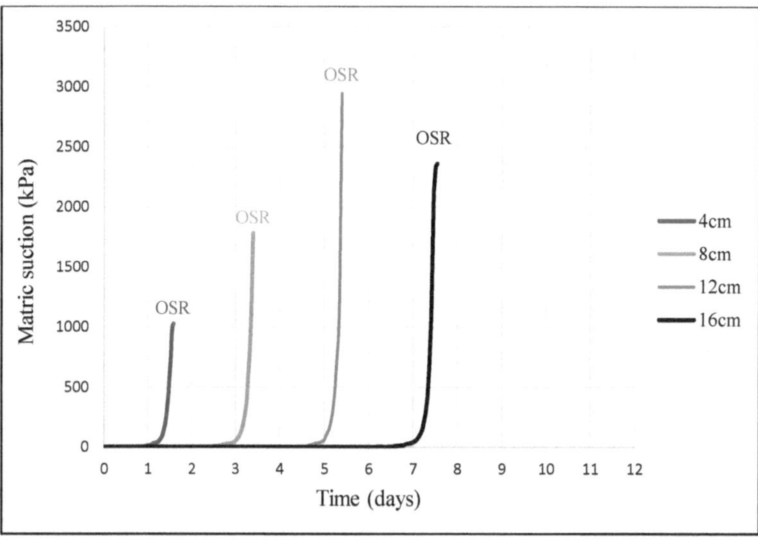

Figure 3.19 Measured Matric Suction Values (Heated-Air flow test)

3.6.2.3 Measured Temperature Values (Heated-Air flow test, constant air temperature in the pipe, T=60°C)

Measured temperature values in this heated air flow test are shown in Figure 3.20. The temperatures measured by the sensors show that the temperatures in the soil near the sensors did not reach the heated air temperature in the pipe even at the location of the sensor located closest to the pipe.

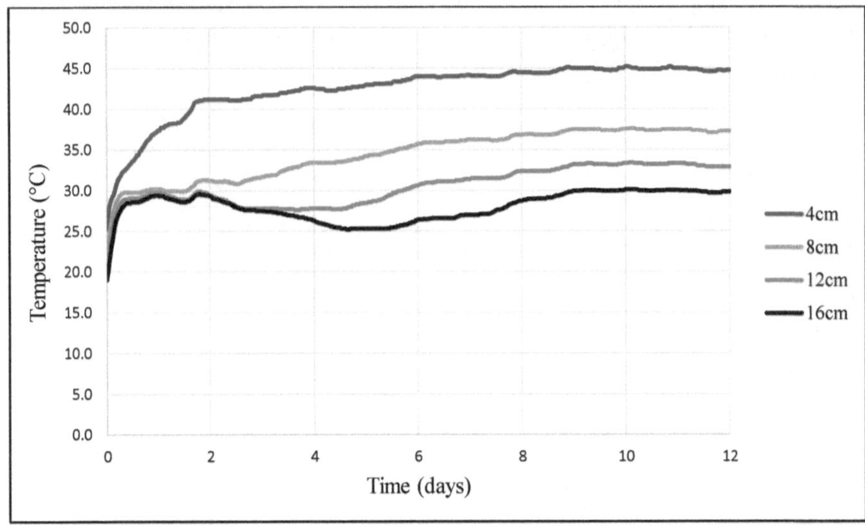

Figure 3.20 Variation of temperature vs time at various depths during the heated air flow test

3.6.3 Test Number #3: Heated-Air flow test (ASL-type #2)

In this heated air flow test, the sensors were placed on a diagonal line in the sand mass as shown in Figure 3.21. It was believed that this new arrangement of sensors would eliminate some of the difficulties caused by the sensors placed on a vertical line (i.e., ASL-type #1). In addition to the changes made to the locations of the sensors, the temperature was not kept constant during the entire duration of the test. Once the preparation of the test setup was completed at room temperature, heated air flow test started at a temperature of 25°C. After 2 days, the temperature of the heated air was increased to 35°C. The test continued at this temperature until 4 days from the beginning of the test. The temperature was then increased to 40°C and kept at that value until the end of the test at day 6.

3.6.3.1 Variation of the VWC at four locations during the heated air flow test (Heated Air temperature varied during the test)

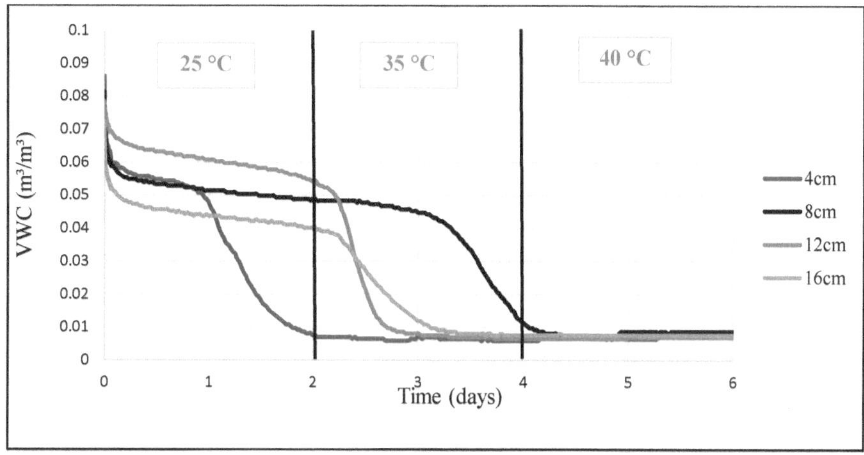

Figure 3.21 Variation of the VWC in heated air flow test. Three different temperatures were used in this test

Figure 3.21 shows that the volumetric water contents at all sensor locations reduced to very small values as a result of heated air flow. Time required for the VWC to reach its smallest value was related to the distance between the sensor and the top of the pipe.

3.6.3.2 Measured matric suction values at four locations in sand during the heated air flow test (Heated air temperature varied during the test)

The measured matric suction values are shown in Figure 3.22. It appears that the possible range of measurements by the sensors at locations 8 cm and 12 cm from the top of the pipe was reached rather early. This does not mean, however, that the matric suction values would not continue to increase after these points. It simply indicates the limitations of the instrumentation.

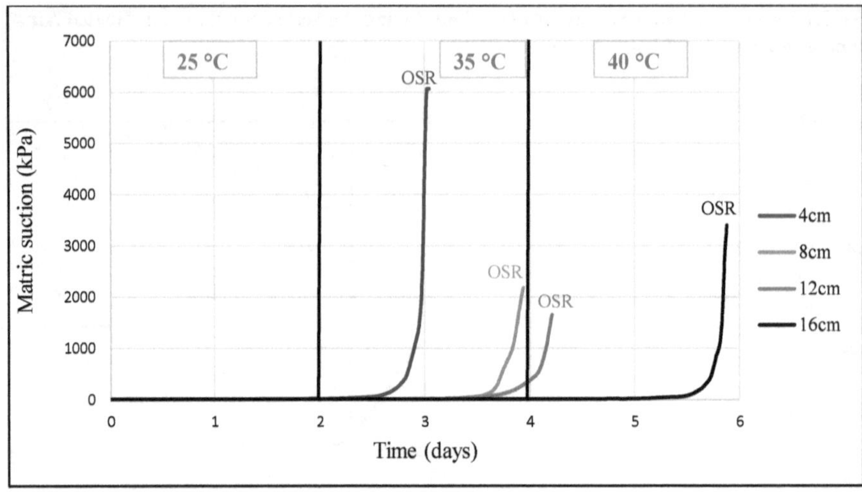

Figure 3.22 Variation of the matric suction in heated air flow test. Three different temperatures were used in this test (Sensors were placed on a diagonal line)

3.6.3.3 Measured temperatures at four locations in sand mass during heated air flow test (Heated air temperature varied during the test)

Measured temperatures in this heated air flow test are shown in Figure 3.23. The temperatures measured by the sensors show that the temperature at the sensor located closest to the pipe is higher than the temperatures measured at other locations as expected.

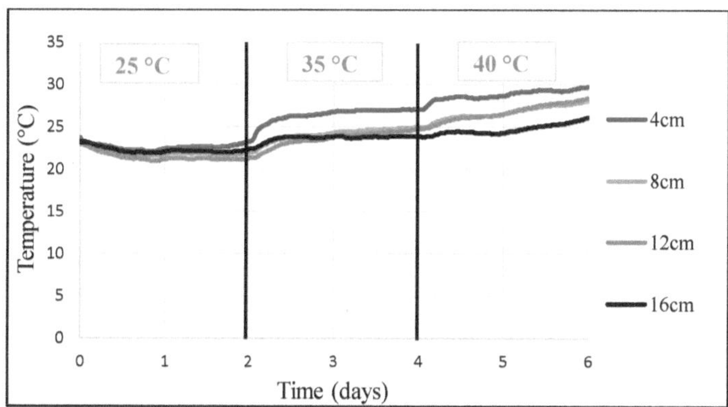

Figure 3.23 Variation of the temperatures in heated air flow test. Three different temperatures were used for the heated air in this test (Sensors were placed on a diagonal line)

3.6.4 Test Number #3: Heated-Air flow test (ASL-type #3)

In this heated air flow test, the sensors were placed on two horizontal lines in the sand mass as shown in Figure 3.12. The differences between this test and Test Number #2 are: (1) the locations of the sensors, (2) the time interval for each temperature was changed, and (3) an additional temperature step (30°C) was included in the experiment. There were four sensors in use to measure the volumetric water contents and four sensors to measure the matric suction. Two of them were located 4 cm away from the pipe. They are labelled as SEN4-1 and SEN4-2. The other two sensors were located 8 cm away from the pipe. They are labelled as SEN8-1 and SEN8-2.

3.6.4.1 Variation of the VWC at four locations during heated air flow test (Heated Air temperature varied during the test)

Figure 3.24 shows the volumetric water content measurements for this test.

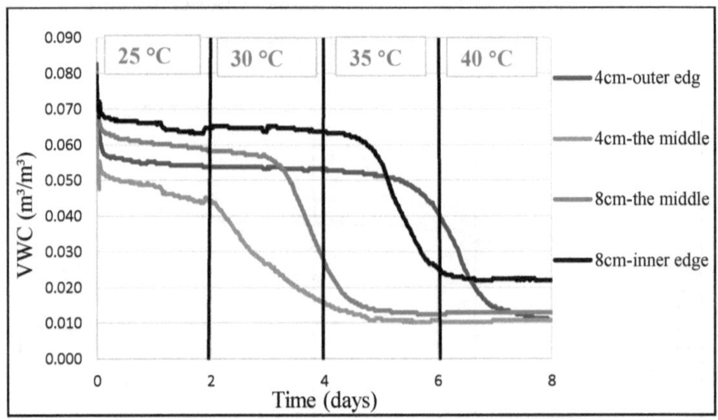

Figure 3.24 The variation of the VWC at four sensor locations

Figure 3.24 shows that the volumetric water contents at all sensor locations reduced to small values as a result of heated air flow. Time required for the VWC to reach its lowest value was related to the distance between the sensor and the top of the pipe as well as how close the sensors were to the center of the sand box.

3.6.4.2 Variation of the matric suction at four locations for heated air flow test (Heated Air temperature varied during the test)

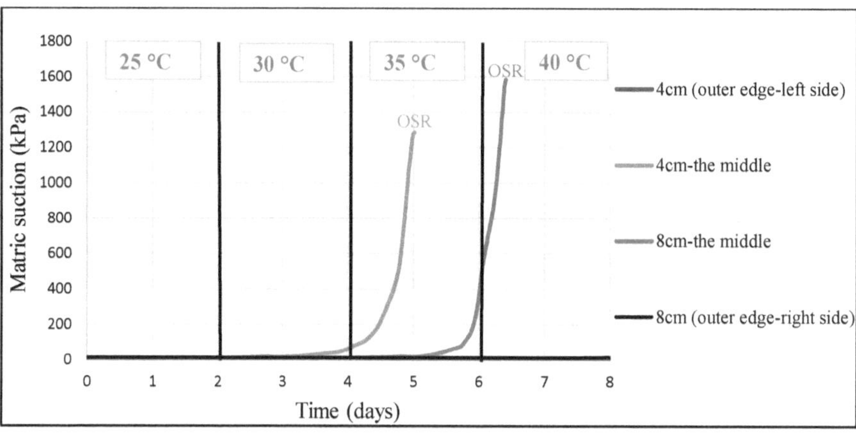

Figure 3.25 Variation of the matric suction in heated air flow test. Four different temperatures were used in this test (Sensors were placed on two parallel lines in the sand mass)

OSR, OSR, OSR or **OSR** = Out of Sensor Range

Figure 3.26 illustrates the drying process in the sand box. These two pictures correspond to the front view of the sand box. The picture on the left was taken two days before the one on the right. The color of the sand inside the enclosed dashed lines is different than the color of the sand outside the dash lines. This means that drying was faster in the middle portion of the sand mass in comparison to the outer edges of the sand box. Thus, the matric suction values were higher at the sensors located in the middle portion of the sand mass.

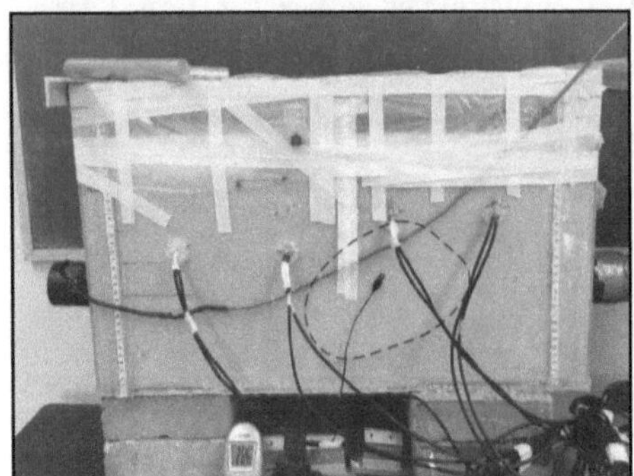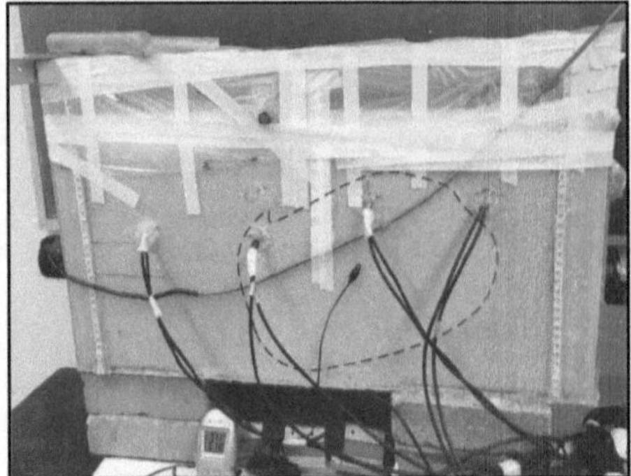

Figure 3.26 The drying process in the sand box

Similar observations were made on the drying process in the sand mass when the side of the sand box was viewed. Dashed lines were marked on the transparent wall of the sand box. Subsequently, the dashed lines were replotted using a graphics software. Figure 3.27 shows the isolines of the variation in drying front with time (in scale) during the test.

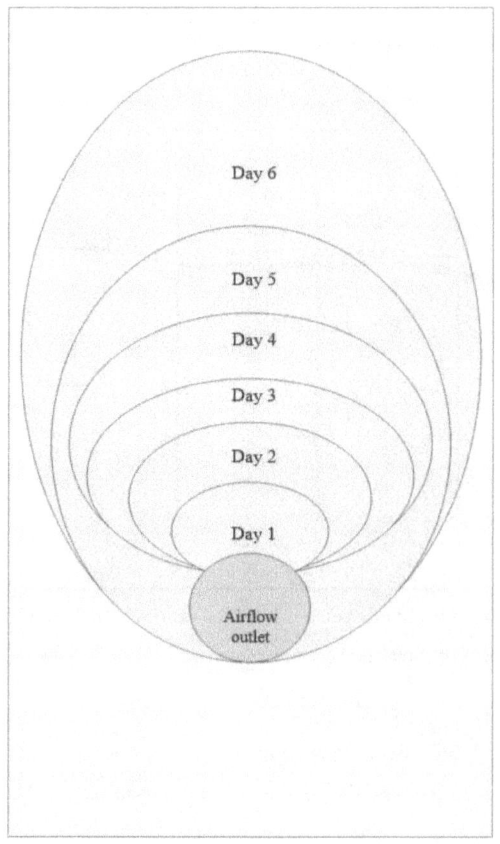

Figure 3.27 Isolines showing the variation in drying front with time

3.6.4.3 Measured temperatures at four locations in sand during heated air flow test (Heated air temperature varied during the test)

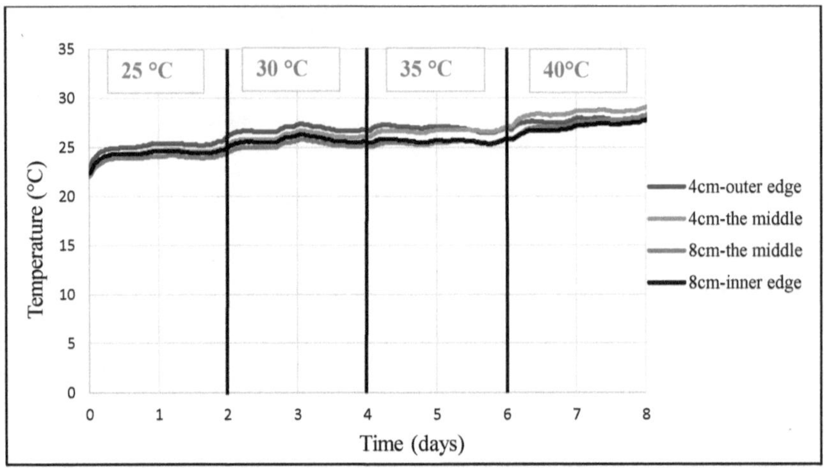

Figure 3.28 Variation of the matric suction in heated air flow test. Four different temperatures were used in this test (Sensors were placed on two horizontal lines in the sand mass)

3.7 Summary

The reduction in the VWC was a function of time as well as the location of the sensors. In general, sensors closer to the pipe showed loss of moisture in the soil higher than the sensors farther away from the pipe for a given period of time. More reliable measurements of VWC values below the residual moisture content values will most likely require different type of sensors than the ones used in these laboratory experiments.

As the matric suction is related to VWC through SWCC, the measured matric suction values became larger as the VWC became lower during heated air flow tests. For very small values of VWC, matric suction could not be measured reliably. It appears that there is a range of measurements with reliable sensor readings. The combined effects of temperature, VWC, and matric suction determine the reliable range of measurements.

Experimental results clearly show that: (1) heated air flow causes a large amount of reduction in VWC in the soil, and (2) matric suction increases as it is very much desired when a capillary barrier is constructed. The increase in matric suction and the decrease in VWC will depend on the temperature of the heated air flow. It would also be reasonable to add that the size of the pipe, the size of the perforations in the pipe wall, the position of the pipe with respect to the interface between the fine grained and coarse grained soil layers would all affect the performance of a capillary barrier and they should be considered in future investigations.

CHAPTER 4

DEVELOPMENT OF A NUMERICAL MODEL FOR LABORATORY EXPERIMENTS

4.1 Introduction

In this chapter, the development of a numerical model for the experimental work presented in Chapter 3 is described. The laboratory experiments demonstrated the changes in the behavior of a coarse grained soil layer under the influence of heated air flow. In the numerical model, the coupled heat and moisture transfer (in both liquid and vapor phases) processes are taken into consideration. Comparisons between the experimental and numerical results are made to investigate the possibility of using the developed numerical model in future simulations of capillary barrier systems, which include both fine and coarse-grained soil layers.

All numerical simulations presented in this chapter were performed using the well-known commercial software, GeoStudio 2018 R2 (GEO-SLOPE International Ltd), which comprises several modules. Two of these modules, namely SEEP/W and TEMP/W were used simultaneously to perform the coupled analyses. SEEP/W models fluid flow problems for both saturated and unsaturated soils, while TEMP/W calculates the heat transfer taking place in the analysis domain. Relevant details of these software applications are provided in the next section.

The air flux was not included in the modelling phase due to its limited effect on the physical model. Since the sand layer was sealed with a plastic film, any small pressure buildup would have caused the plastic film to inflate. This was observed if the tube through which air flows was obstructed (Figure 7.3). For field applications, it is also expected that air flow will have minimal impact on the desaturation of the sand layer in its intended application within a capillary barrier, since at such time when a significant amount of water would be present in the sand, it is expected that the overlaying clay would be saturated. As such, the air pressure within the system would have to exceed the clay layer's air entry value to achieve any significant desaturation through that path. Any buildup of pressure would instead serve to drive the air through the pipe, thus reducing pressure and carrying moisture on the way out. It was therefore felt that the addition of that flow path, while more exact physically would not provide a sufficient means to desaturate the system to justify its inclusion in the simulation.

4.2 Theoretical aspects of the numerical analysis using SEEP/W and TEMP/W

When used jointly, the SEEP/W and TEMP/W programs (GeoStudio 2018 R2) have the ability to couple heat and water transfer (in liquid and vapor phase) in saturated and unsaturated soils. The heat transfer mechanism in the analysis includes heat transfer by conduction, convection, and forced heat convection through water in both phases (i.e., liquid and vapor). The numerical analysis of the coupled processes provides the temperature distribution, the volumetric water content and the matric suction within the soil profile as a function of time.

The governing principles for two-dimensional analysis of water flow and heat transfer are expressed by Equations 4.1 and 4.2, respectively.

Equation 4.1 is the mass transfer equation and it is a direct derivative of Richard's equation for transient flow in unsaturated soils (GEO-SLOPE International Ltd, 2014). It postulates that, within the interior of the soil mass, the difference between liquid or vapor flux that is entering the representative elemental volume (REV) of soil and leaving it during the same time interval is equal to the change in the volumetric water content of REV. If the REV is located at the boundary between the soil mass and the atmosphere, the applied boundary flux has to be added to the equation, as shown below.

$$\frac{1}{\rho}\frac{\partial}{\partial x}\left[D_v\frac{\partial P_v}{\partial x}\right] + \frac{1}{\rho}\frac{\partial}{\partial y}\left[D_v\frac{\partial P_v}{\partial y}\right] + \frac{\partial}{\partial x}\left[K_x\frac{\partial\left[\frac{P}{\rho g}+y\right]}{\partial x}\right] + \frac{\partial}{\partial y}\left[K_y\frac{\partial\left[\frac{P}{\rho g}+y\right]}{\partial y}\right] + Q = \lambda\frac{\partial P}{\partial t}$$

(4.1)

where P is pressure, P_v is vapor pressure of soil moisture, K_x is hydraulic conductivity in the x-direction, K_y is hydraulic conductivity in the y-direction, Q is applied boundary flux, D_v is vapor diffusion coefficient, y is elevation head, ρ is density of water, g is acceleration due to gravity, λ is capacity for heat storage, and t is time.

Equation 4.2 is written as a function of pressure instead of total head as a driving force in order to couple the heat and mass transfer processes. The heat transfer equation unitized by the software is a modified Fourier equation, which was originally written for conductive heat transfer. The modification to the Fourier equation includes the vapor transfer in addition to convective heat transfer caused by water flow.

$$L_v \frac{\partial}{\partial x}\left[D_v \frac{\partial P_v}{\partial x}\right] + L_v \frac{\partial}{\partial y}\left[D_v \frac{\partial P_v}{\partial y}\right] + \frac{\partial}{\partial x}\left[K_{tx}\frac{\partial T}{\partial x}\right] + \frac{\partial}{\partial y}\left[K_{ty}\frac{\partial T}{\partial y}\right] + Q_t + \rho c V_x \frac{\partial T}{\partial x} + \rho c V_y \frac{\partial T}{\partial y}$$
$$= \lambda_t \frac{\partial T}{\partial t}$$

$$(4.2)$$

where ρc is volumetric specific heat value, K_{tx} *is* thermal conductivity in the x-direction, K_{ty} is thermal conductivity in the y-direction and assumed equal to K_{tx}, V_{xy} is the Darcy water velocity in x and y directions, Q_t *is* applied thermal boundary flux, and L_v *is* latent heat of vaporization. Two heat transfer mechanisms were considered in the current simulation, namely, heat transfer by conduction and heat transfer by convection (heat flow in unsaturated soils). The heat conduction takes place when there is a temperature gradient within the soil profile. Accordingly, the heat is transferred from a region with high temperature to a region with low temperature. In contrast, the heat transfer by convection takes place in the case of water vapor moving through the soil mass (Anderson 2005).

The heat and mass (in liquid and vapor phases) transport equations are dependent on three unknown variables, namely pressure (P), temperature (T), and vapor pressure (P_v). Equation 4.3 presents the relationship between these variables.

$$P_v = P_{vs}\left(e^{\frac{-P.w}{\rho.R.T}}\right) = P_{vs}h_{r\ air}$$

$$(4.3)$$

where P_{vs} is saturated vapor pressure of pure free water, w is molecular mass of water vapor, R is universal gas constant, T is temperature (K), h_r is relative humidity of air.

4.3 Description of the numerical model for the simulation of the laboratory experiments

In order to use the Geostudio software for coupled analysis, the following material properties have to be provided as input.

    a- Hydraulic properties of the sand.

    b- Thermal properties of the sand.

4.3.1 Functions for the hydraulic properties of the sand

Hydraulic properties of the unsaturated sand to be used in the numerical analysis include the soil-water characteristics curve and the hydraulic conductivity as a function of matric suction. The equations developed by Fredlund and Xing (1994) for SWCC were used in the numerical simulations. Figure 4.1 shows the SWCC used in the analysis.

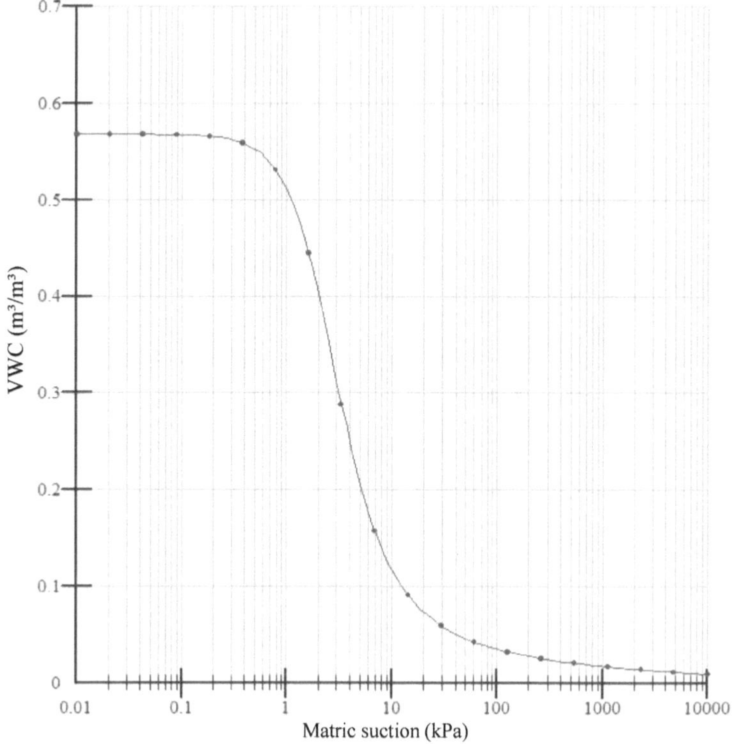

Figure 4.1 SWCC used in the numerical analysis

The numerical analysis also requires, as an input, the relationship between the unsaturated hydraulic conductivity and matric suction. Figures 4.2 and 4.3 show the hydraulic functions for heated-air flow test (ASL-type #1), and heated-air flow test (ASL-type #2), respectively.

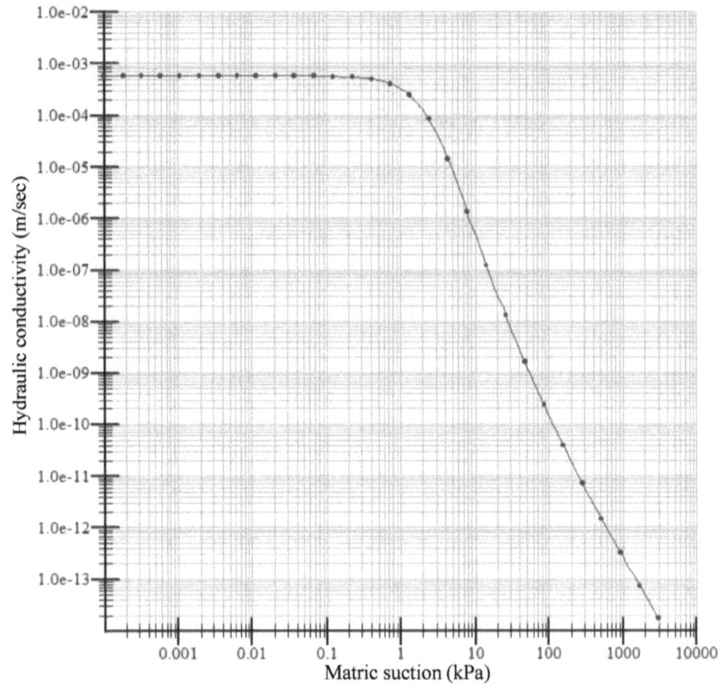

Figure 4.2 Hydraulic conductivity versus matric suction relationship used in the numerical analysis (ASL-type #1)

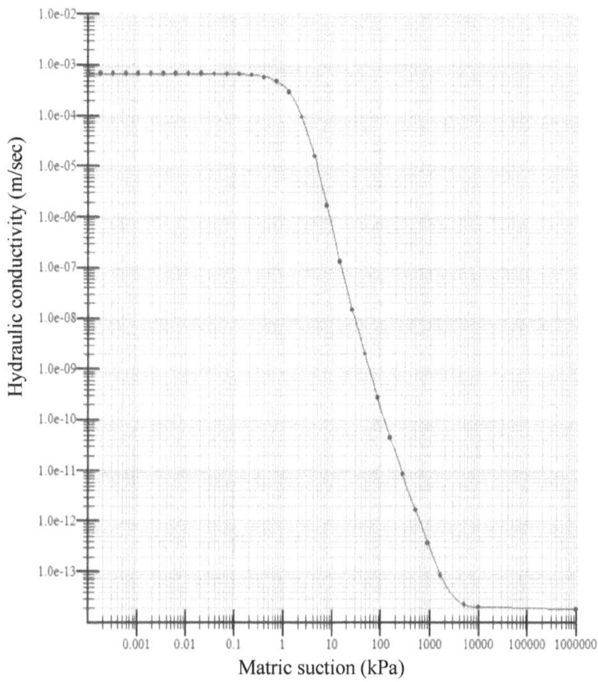

Figure 4.3 Hydraulic conductivity versus matric suction relationship used in the numerical analysis (ASL-type #2)

4.3.2 Functions for the thermal properties of the sand

The heat transfer in soils is mainly a function of the volumetric water content and the temperature gradient. The input for the numerical analysis requires: (a) thermal conductivity versus volumetric water content relationship (Figure 4.4), and specific heat capacity as a function of volumetric water content (Figure 4.5). These input relationships/functions also depend on the soil type.

The activation temperature (the initial temperature) is entered as 24°C.

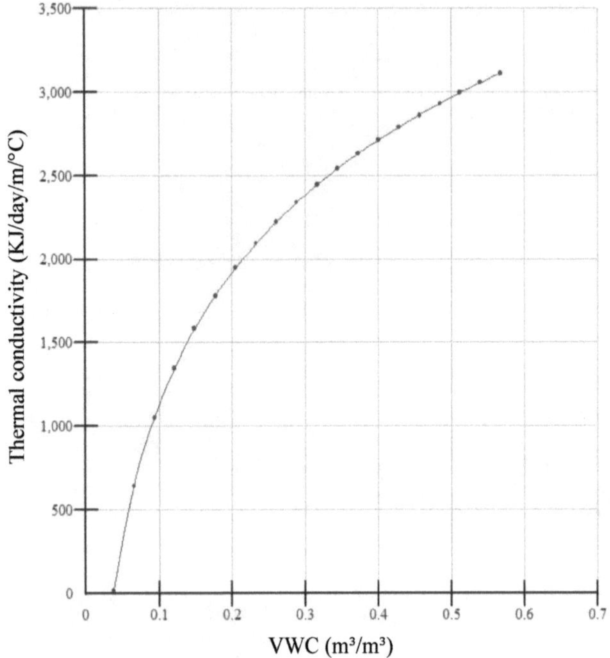

Figure 4.4 Thermal conductivity vs. VWC relationship

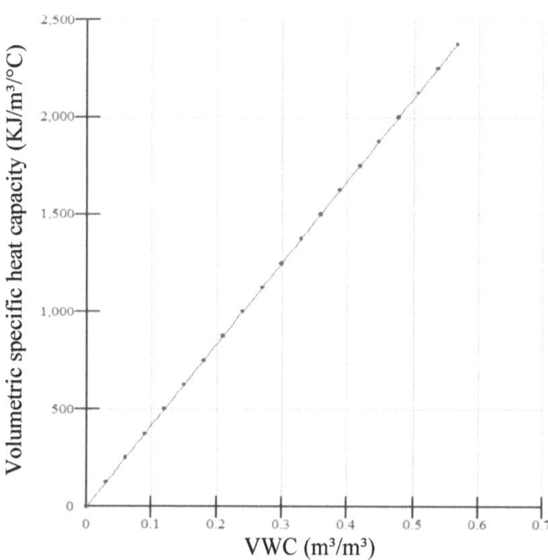

Figure 4.5 Volumetric specific heat capacity versus volumetric water content relationship

4.4 Heated-Air flow test (ASL-type #1)

A model size of 0.2 m by 0.2 m was used to simulate the test with constant temperature of 60°C. Approximate Upside-Down Analysis (AUDA) represents what is happening between the perforation hole and the top of the soil sample as shown in Figure 4.6.

The 2D and Axisymmetric model of the sand box were previously used to simulate the test but both suffered from some limitations. For example, it was not possible to define the land-climate interaction boundary condition on the pipe located inside the domain (i.e., to consider the relative humidity and the air speed at the pipe-sand interface to solve the numerical problem). This boundary condition is very important due to its effect on the water removed from the model through the pipe only. The current approach (i.e., AUDA) provided close solution to the target numerical solution (i.e., the simulated values are closer to the measured values). Meanwhile, the minimum VWC was not reached using other simulation methods.

4.4.1 The initial conditions

All transient analyses must begin with known (initial) values of the variables at time zero. The numerical analysis of the laboratory experiments started at room temperature of 24°C before the application of the heated air flow. A steady state analysis was conducted in order to establish the initial condition values for the matric suction (i.e., the activation pore-water pressure).

4.4.2 The boundary conditions

In finite element analysis, after the initial conditions have been provided as an input, the coupled balance equations are solved as a function of time to satisfy the boundary conditions. In the current numerical analysis of the laboratory experiments, the boundary conditions included: i) hydraulic boundary conditions, and ii) thermal boundary conditions. Table 4.1 summarizes both sets of boundary conditions.

Two simulation models were initially developed to examine the effect of boundary condition selection at the pipe-soil interface on the predicted volumetric water content and matric suction. The first model represented the pipe perforations as multiple openings, while the second model represented the perforations as a continuous straight slot, as shown in Figure 4.6. In both cases, the boundary was modelled as land-climate interaction, mesh size, and time step were kept the same. The second model predictions were found to agree better with experimental data.

4.4.3 Hydraulic boundary conditions

The analysis domain and the hydraulic boundary conditions are shown in Figure 4.6. The zero water flux boundary condition was specified on the left side, the right side and the bottom of the sand box. The land-climate interaction boundary condition was applied on the top surface of the model using different climate functions (eg: relative humidity, air speed, and temperature as a function of time).

4.4.4 Thermal boundary conditions

Zero heat flux boundary condition was defined on the left side, the right side and the bottom of the sand box.

Both heat convection, caused by vapor transfer in the soil profile, and heat conduction were considered during the analysis.

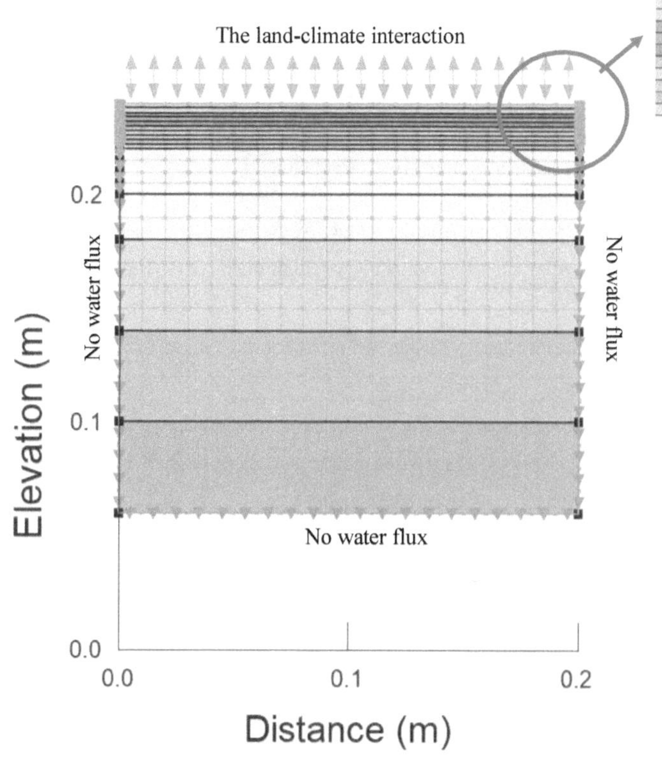

Figure 4.6 Hydraulic and thermal boundary conditions for the sand box

Table 4.1 The boundary conditions

Color	Name	Category	Kind	Parameters
	climate data	Hydraulic	Land-Climate Interaction	
	zero heat flux	Thermal	Heat Flux	0 kJ/d/m²
	zero water flux	Hydraulic	Water Flux	0 m/d

4.4.5 Mesh size and time steps

In general, mesh size and time steps used in a transient analysis can affect the finite element results. In the present study, the time step and the mesh size (i.e. coarse or fine) were varied until the ones yielding results that best matched the measured experimental values were found. These were utilized in all subsequent numerical simulations.

As shown in Figure 4.6, quads and triangles were used for the shape of the elements with a default element size of 0.01m. A rectangular grid of quads of 0.01m was used for the top surface layer.

The mesh at the surface was simulated using the "surface layer" (more details about the surface layer are provided in Chapter 6).

A total of 940 elements were used to simulate the model of the sand box (0.20 m in height, 0.20 m in depth). The analysis was carried out to simulate a test duration of 12 days. The time increment of 0.083 days is used during the analysis.

4.4.6 Results and discussion

The results of the finite element analysis for heated-air flow test (ASL-type #1) are presented in terms of the changes in VWC and matric suction.

4.4.6.1 Volumetric water content changes over time

Figure 4.7 shows the calculated values of the VWC at four different locations in the sand mass. As evident from the figure, there is a good agreement. It is not only the VWC values but also the corresponding time for the drying process is simulated quite well.

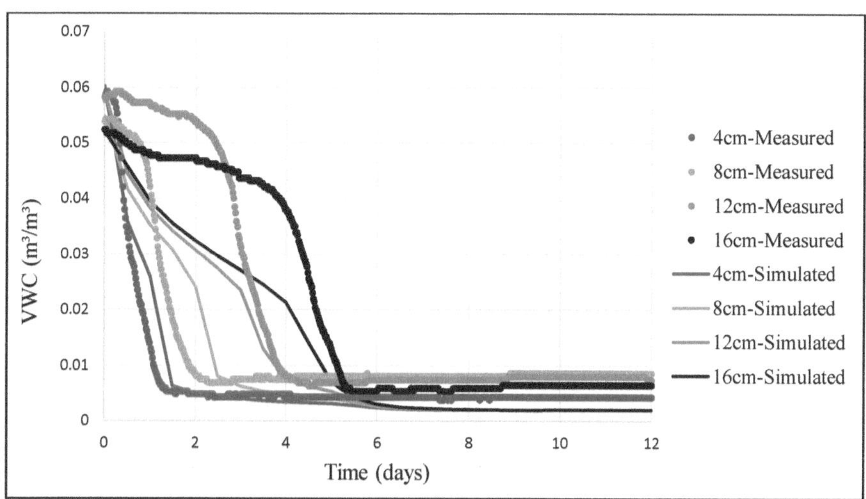

Figure 4.7 The comparison between the measured and simulated VWC

It can be seen from Figure 4.7 that the measured volumetric water content data at 4 cm, 8 cm, 12 cm and 16 cm away from the pipe are generally in good agreement with the simulated values for the same locations. However, there is a significant variation between the measured and the simulated water contents at early times. It can be noted that the trends at 4 cm and 8 cm are better simulated, while 12 cm and 16 cm show greater differences between simulated and measured values, especially at early time of the test. Yet, with time these trends too come close to the measured values, that is at day 4 for location of 12 cm away from the pipe, and at day 5.5 for location of 16 cm. This behavior can be explained with the fact that the 4 cm and 8 cm locations are closer to the defined boundary, while the 12 cm and 16cm are farther away and thus, more sensitive to minor variations in temperature, moisture and density. Any small variation in these parameters' input values might cause a significant departure of the predicted volumetric water content values from the experimental results.

4.4.6.2 Pore water pressure changes over time

Figure 4.8 shows the calculated values of matric suction at four different locations in the sand mass. It can be seen from Figure 4.8 that the calculated matric suction values are consistent with the volumetric water content values shown in Figure 4.7. However, it was not possible to make comparisons between the calculated matric suction values and matric suction values measured in laboratory experiments. Instrumentation in laboratory experiments was not able to read high matric suction values at conditions of very low VWC combined with very high temperatures.

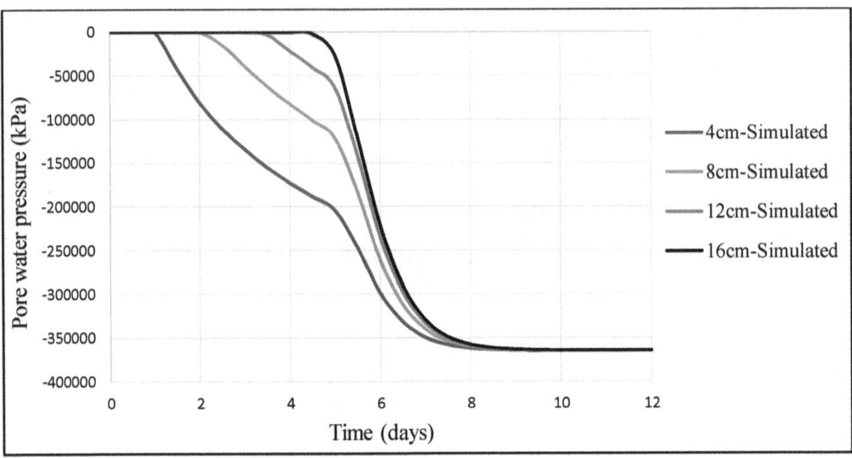

Figure 4.8 Pore water pressure variation versus time

4.5 Heated-Air flow test (ASL-type #2)

The simulation was carried out for different temperature values namely, 25°C, 35°C and 40°C. The "No water flux" boundary condition was applied on the sides and on the bottom of the model. The "No heat flux" boundary was applied on the sides and on the bottom of the model. The "land-climate interaction" boundary condition was applied on the top surface of the model using different climate functions (eg, relative humidity and temperature as a function of time).

4.5.1 Mesh size and time step

Different element size and mesh densities were used in the simulation to improve the solution accuracy. A rectangular grids of quads (0.005 m) were applied on the model. A total of 2440 elements were created to analyze the FE problem.

The analysis was carried out for a total duration of 6 days. The time steps were selected on the basis of element size and hydraulic conductivity values as 0.0138 day. Figure 4.9 shows the model domain for the analysis. Table 4.2 describes the applied boundary conditions (hydraulic and thermal boundary conditions).

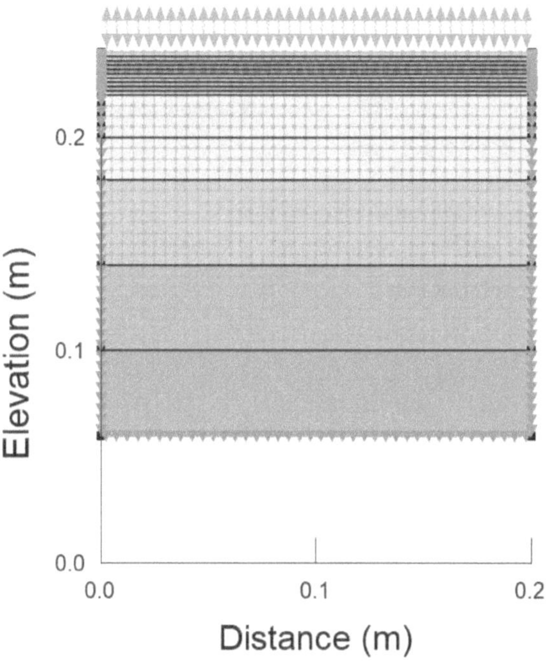

Figure 4.9 Hydraulic and thermal boundary conditions for the sand box

Table 4.2 The boundary conditions

Color	Name	Category	Kind	Parameters
	climate data	Hydraulic	Land-Climate Interaction	
	zero heat flux	Thermal	Heat Flux	0 kJ/d/m²
	zero water flux	Hydraulic	Water Flux	0 m/d

4.5.2 Results and discussion

The results of the finite element analysis for heated-air flow test (ASL-type #2) are presented in terms of the changes in VWC and matric suction.

4.5.2.1 Volumetric water content changes over time

Figure 4.10 shows the calculated values of VWC at four different locations in the sand mass for heated-air flow test (ASL-type #2).

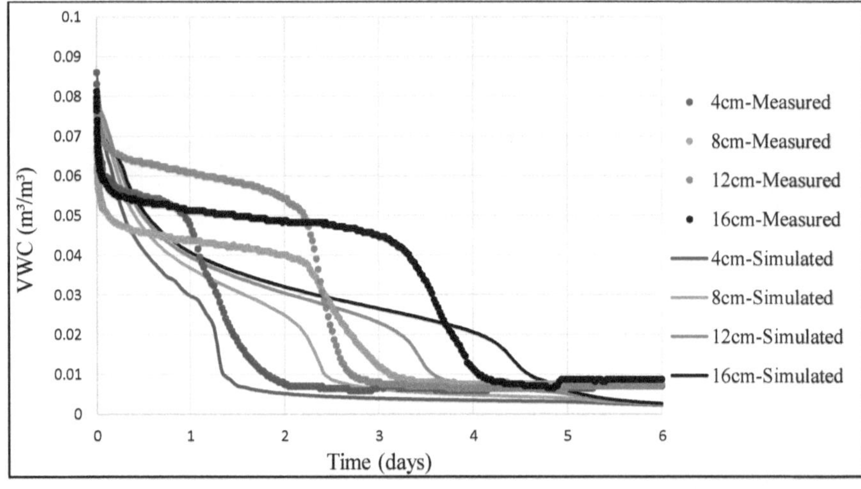

Figure 4.10 The comparison between the measured and simulated VWC

A comparison indicates that the simulated values of VWC at 4 cm, 8 cm, 12 cm and 16 cm are close to the measured values. This difference (the variation) between the two trends are due to the input values used in the analysis, taking into the account the sensor accuracy limitations for measurements of temperature, RH, wind speed (airflow speed in the pipe), whereas these factors play a very important role in the calculated values (they are input parameters for the defined boundary conditions).

4.5.2.2 Pore water pressure changes over time

Figure 4.11 shows the calculated values of matric suction at four different locations in the sand mass for heated-air flow test (ASL-type #2). It can be seen from Figure 4.11 that the calculated matric suction values are consistent with the volumetric water content values shown in Figure 4.10.

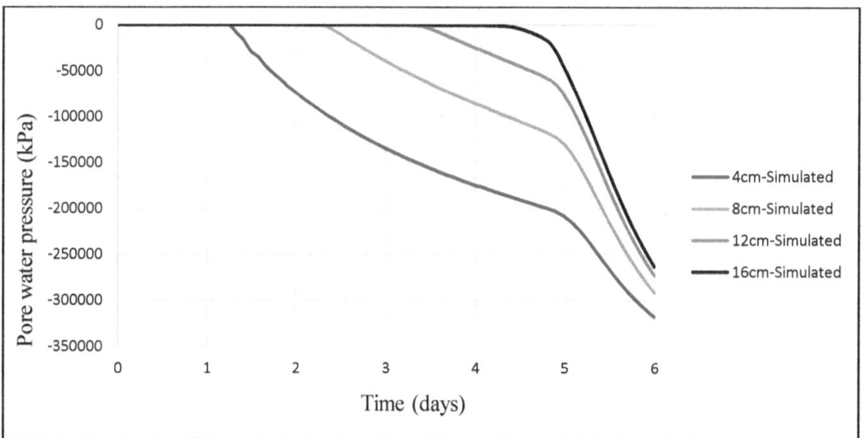

Figure 4.11 Pore water pressure variation versus time

4.6 Sensitivity analysis

It should be noted that the Soil Water Characteristics Curve (SWCC) is one of the most important components in the numerical analysis. In the present study, the experimental data points shown in Figure 4.5 have been obtained from laboratory tests. For the purpose of the numerical analysis, the Fredlund and Xing (1994) model was chosen to represent the experimental data. A sensitivity analysis was carried out to determine the effect of variations in the parameters of Fredlund and Xing (1994) model on the results of the numerical analysis of heated air flow tests in the sand box. The reasons for possible variations in the model parameters include the factors discussed below.

A given type of soil at different densities will have different SWCCs. The density of the sand at different parts of the sand box is not necessarily identical to the density of the sand used in the laboratory experiment to define the SWCC. The method of sand placement in the sandbox can easily produce some variations in soil density from location to location. More importantly, the presence of the sensors in the sand box to measure the VWC and suction is a significant contributor to the variations in the density of sand. In addition, a different magnitude of temperature changes in different parts of the sand box during the heated air flow tests can cause further variations in the density of the soil. Furthermore, very high suction values generated in the heated air flow tests can also cause variations in the soil density in the sand box. In the laboratory experiment to determine SWCC, the maximum matric suction measured was about 6.5 kPa whereas in the heated air flow tests, the matric suction values were in the order of hundreds of thousands of kPa. The variation was due to the nature and limitations of the laboratory test procedure used to determine SWCC. Moreover, the wetting and drying cycles of many soils result in different SWCCs (i.e., a hysteresis effect).

In order to account for the effect of variations in the model parameters on the results of numerical simulations, a sensitivity analysis is conducted. There are three fitting parameters in the Fredlund and Xing (1994) model. These are α, n, and m. Firstly, upper and lower bound values were defined for each parameter. Next, a separate analysis where one parameter is assigned values within its upper and lower bounds, while the other parameters were kept constant. The upper and lower bound values were chosen to be at 25% of the measured values given in Table 4.3 (different percentage can be used to represent the upper and lower bound values). More sophisticated sensitivity analysis is left for future studies. However, the sensitivity analysis conducted here helps to explain why some of the experimental data determined by the sensors in the heated air flow tests are not exactly the same as determined by the numerical analyses (i.e. how the model can be effected by any variation in the input parameters).

Table 4.3 Fitting parameters calculations

α	n	m	Magnitude
2.1371	1.9494	1.3784	Original fitting parameters
2.671375	2.43675	1.723	+25% of original
1.602825	1.46205	1.0338	-25% of original

4.6.1 Results and discussion

In this section, the results of the finite element analysis using different fitting parameters are presented in terms of the changes in VWC.

4.6.1.1 Heated-air flow test (ASL-type #1)

Figures 4.12 through 4.14 show the sensitivity analysis result for heated-air flow test (ASL-type #1).

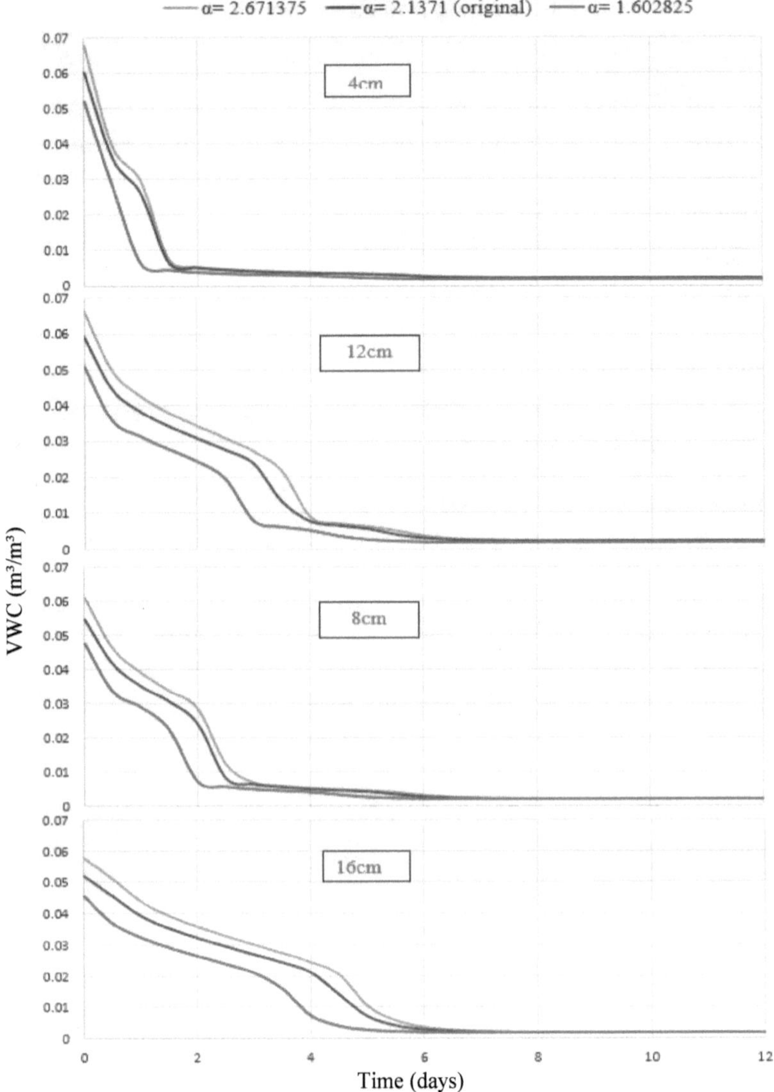

Figure 4.12 VWC plot versus time with n = 1.9494 and m = 1.3784 (α varies)

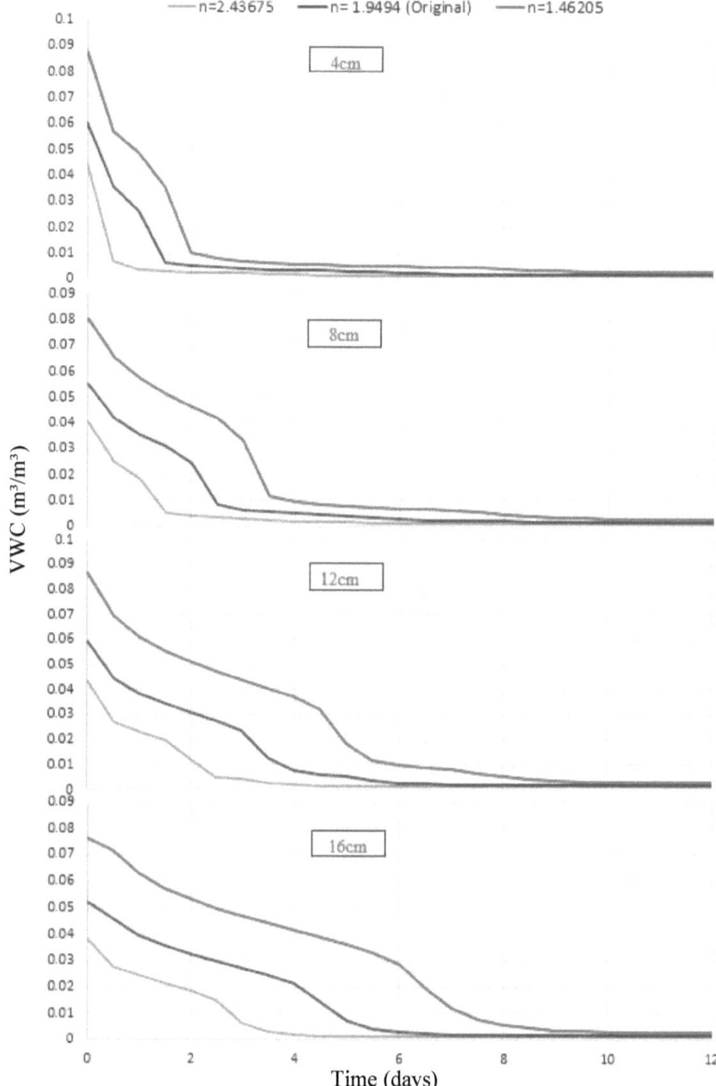

Figure 4.13 VWC plot versus time with $\alpha = 2.1371$ and $m = 1.3784$ (n varies)

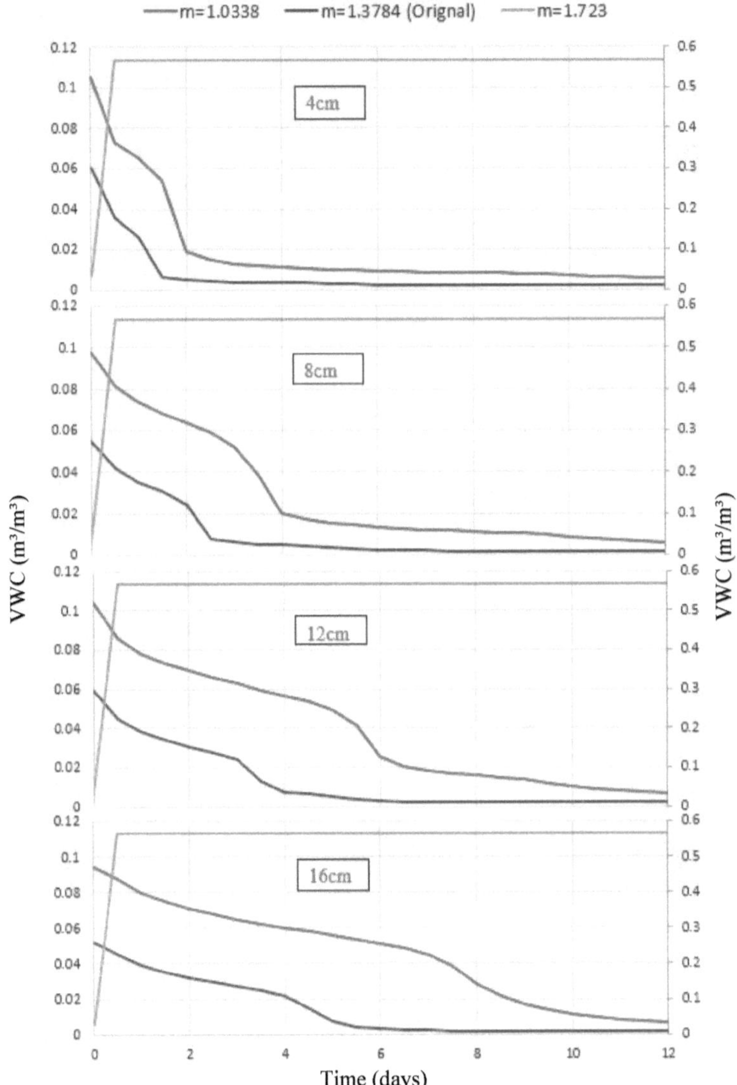

Figure 4.14 VWC plot versus time with $\alpha = 2.1371$ and $n = 1.9494$ (m varies)

4.6.1.2 Heated-Air flow test (ASL-type #2)

Figures 4.15-4.17 show the sensitivity analysis result heated-air flow test (ASL-type #2).

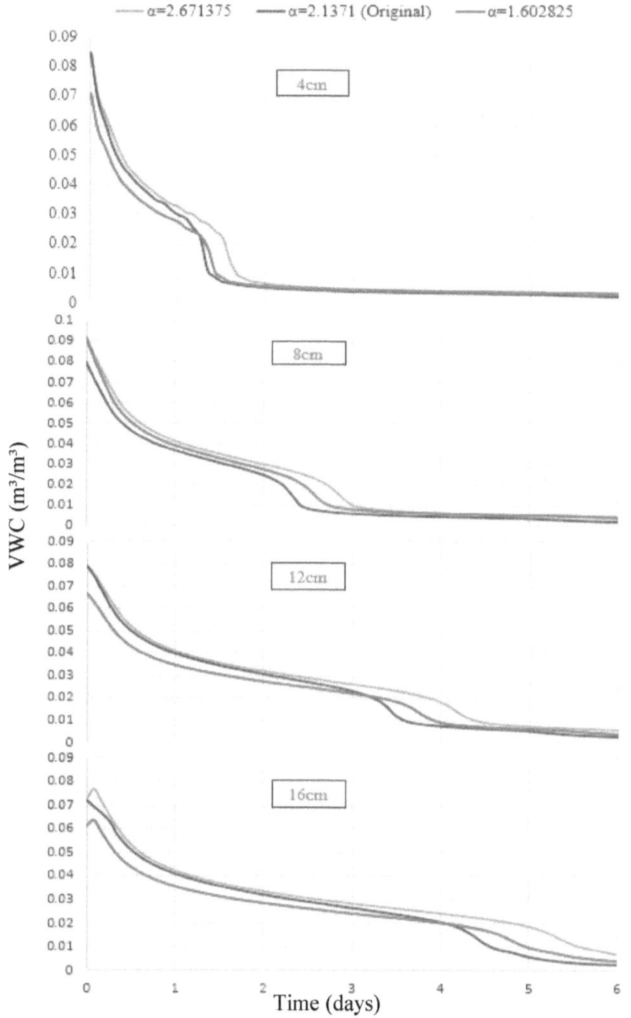

Figure 4.15 VWC plot versus time with n = 1.9494 and m = 1.3784 (α varies)

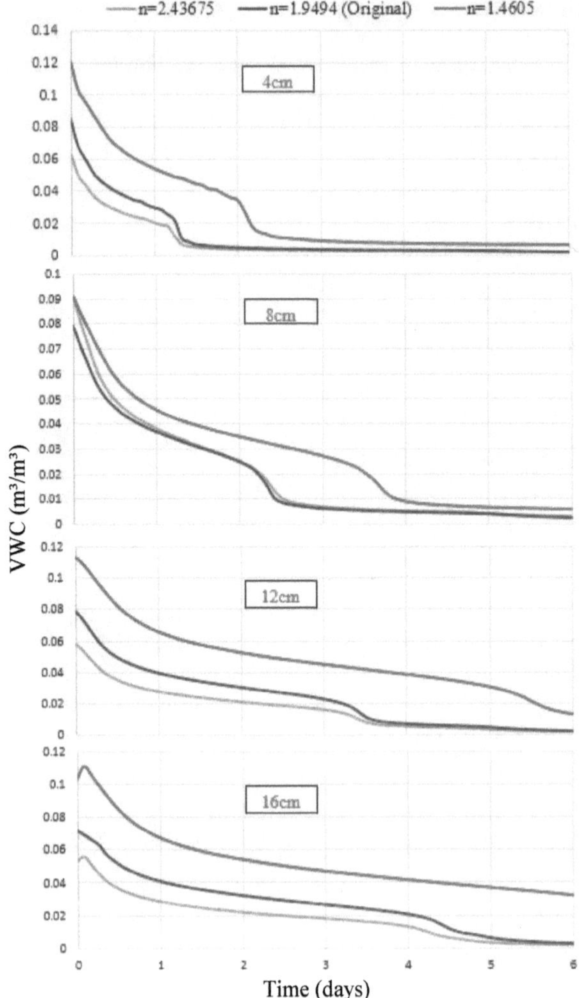

Figure 4.16 VWC plot versus time with $\alpha = 2.1371$ and $m = 1.3784$ (n varies)

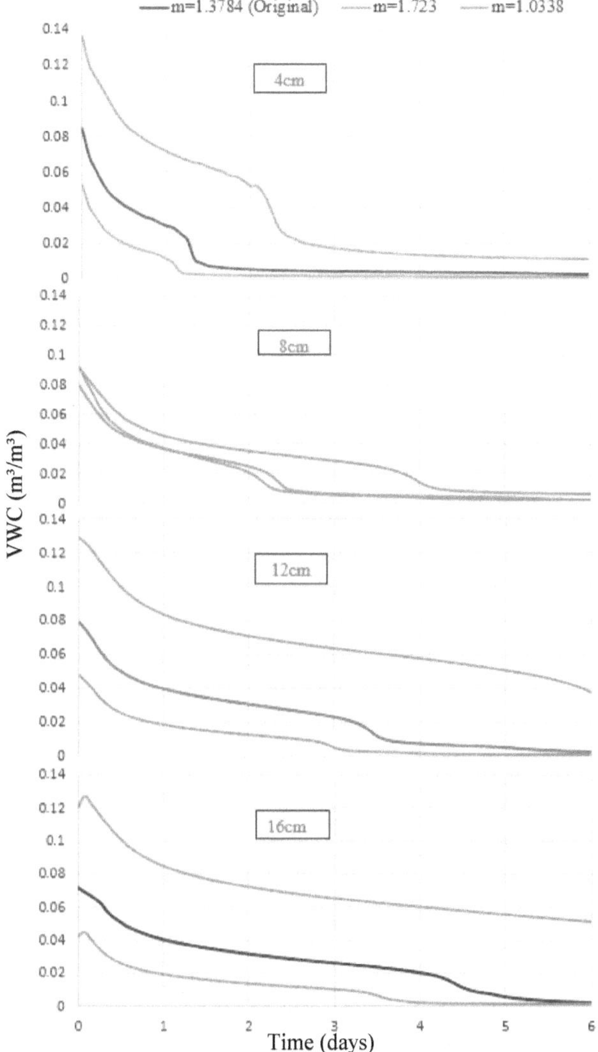

Figure 4.17 VWC plot versus time with $\alpha = 2.1371$ and $n = 1.9494$ (m varies)

4.7 Summary

This chapter presented the numerical model developed for the simulation of the experimental work. Similarly, to the lab experiments, the finite element analysis showed the changes in the behavior of the sand mass under the influence of heated air flow. That behavior was investigated not only at a constant temperature (60°C) but also at different temperature increments. In the numerical model, the coupled heat and moisture transfer (in both liquid and vapor phase) processes were taken into consideration. The trends of the simulated VWC at 12 cm and 16 cm show a delay in the drying process in comparison to the measured values for the same locations. This variation is because the locations of 12 cm and 16 cm are away from the defined boundary conditions in the numerical model.

However, the simulation is dependent on many sensitive variables and estimation of several nonlinear unsaturated soil property functions.

It was not possible to make a comparison between the calculated matric suction values and matric suction values from the laboratory experiments as the instrumentation used in the latter was not capable of measuring the very high matric suction and temperature values.

It was found that the numerical model represented the physical processes reasonably well. Therefore, the author conclude that it is possible to use the procedures developed in the present numerical model in the future simulations of capillary barrier systems which include both fine and coarse grained soil layers.

It can be noted that the model is very sensitive to changes in the SWCC fitting parameters. In particular, the volumetric water content was affected significantly by the choice of fitting parameters and that effect extended to the initial volumetric water content.

It can be also noted that the analysis for the ASL-type #2 is more sensitive to the change in the m- value in comparison with other changes in α and n values.

CHAPTER 5

CASE STUDY (1): AN INVESTIGATION TO DETERMINE THE EFFECT OF HEATED AIR FLOW ON THE PERFORMANCE OF A MODEL SCALE SLOPING CAPILLARY BARRIER

5.1 Introduction

In this chapter, the first of the case studies, CASE STUDY (1), is presented. Tami et al. 2004 (co-authored with Rahardjo, Leong, and Fredlund) reported the results of a model scale laboratory tests on a sloping capillary barrier. Their paper provides detailed information about the design of the model scale experiments, detailed information about the layers of unsaturated soils, the initial conditions, boundary conditions, and the measured data recorded during the entire duration of their experiments. There was no forced air flow or temperature changes during these experiments as will be explained in the next section. In this study, only one of their laboratory tests is used. The details of this particular test will be given in the following paragraphs.

The purpose of this part of the research program was to simulate the effect of heated air flow through the coarse-grained soil layer of the model scale test of Tami et al. (2004) on the performance of this sloping capillary barrier. To achieve this objective, two numerical analyses were conducted. The first analysis was done to simulate the behaviour of the model scale capillary barrier at room temperature and no heated air flow. In the second analysis, heated air flow was added into the numerical model. The comparison of the results of these analyses was done by plotting: (a) the volumetric water contents; (b) matric suction values, and (c) time to breakthrough.

5.2 Physical Model of Sloping Capillary Barrier (Tami et al. 2004)

Figure 5.1 shows a sketch of the physical model (i.e., laboratory scale model). The dimensions and some of the boundary conditions are shown in the figure. This physical model is relatively large although it is a lab scale model. The design of this model, as described by Tami et al. (2004), is based on the unsaturated material properties of several types of soil and a large number of finite element simulations using SEEP/W (GEO-SLOPE International, Ltd. 2000) and SVFlux (Soil-Vision Systems Ltd. 2001).

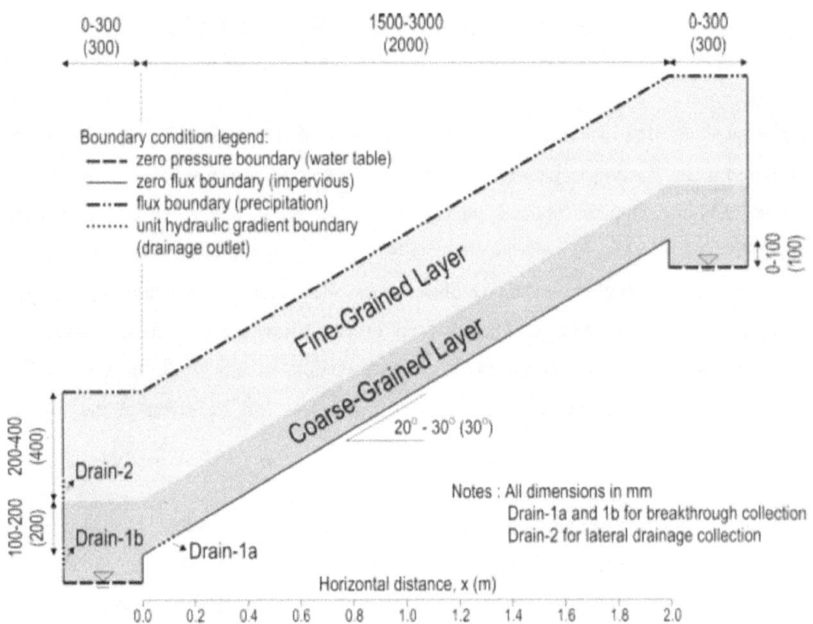

Figure 5.1 Geometry, dimensions, and boundary conditions of the capillary barrier model (from Tami et al. 2004)

In the particular experiment that was analysed in the present study, concrete sand was used as a fine-grained layer and pea gravel was used as a coarse-grained layer (Tami et al. 2004).

The details of the material properties, initial conditions, and boundary conditions are given below.

5.2.1 Types of soils and description of material properties

This two-layer capillary barrier system consisted of a layer of fine-grained soil (concrete sand) at the top and a layer of coarse-grained soil (pea gravel) at the bottom. Based on the Unified Soil Classification System (USCS), the concrete sand was classified as a poorly graded sand (USCS group SP) with less than 1% fines (<0.075 mm). Pea gravel was uniform in grain size distribution and more than 99% of the pea gravel had particle size between 4 mm and 20 mm (USCS group GP). The information provided below is taken from the publication of Tami et al. (2004) and is used in the present study to conduct numerical analysis.

Soil-water characteristics curves of the soils used in the physical model tests are shown in Figure 5.2. Only the curves for the pea gravel and concrete sand are relevant to the present study.

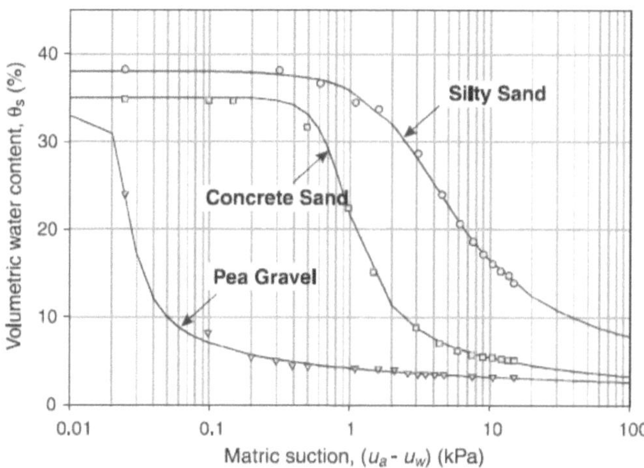

Figure 5.2 Soil-water characteristic curves of the soils used in the design of the capillary barrier system. Experimental data are shown as symbols and the fitted curves are shown as solid lines (from Tami et al. 2004)

The solid lines in Figure 5.2 are obtained by Tami et al. 2004 using the equation of Fredlund and Xing (1994). The parameters used to generate the fitted curves are given in the following table.

Table 5.1 Hydraulic properties of the materials used in designing the capillary barrier physical models (from Tami et al. 2004)

Soil type	k_s (m/s)	θ_s	ψ_a (kPa)	θ_r	ψ_w (kPa)	a (kPa)	n	m
						Best-fit parameters		
Silty sand	3.0×10^{-6}	0.38	0.02	0.11	14.00	2.86	2.23	0.77
Concrete sand	2.4×10^{-4}	0.35	0.50	0.06	1.30	0.64	3.33	0.80
Pea gravel	1.3×10^{-1}	0.33	1.10	0.05	0.05	0.03	3.04	1.03

The definitions of some useful terms related to SWCCs obtained from wetting and drying tests are given next. The air-entry value, ψ_a is defined as the matric suction at which air first enters the saturated pores of the soil along the drying path. On the other hand, the water-entry value, ψ_w is defined as the matric suction at which water first enters the dry pores of the soil in the wetting path, and its value was estimated from the inflection point on the SWCC (Figure 5.3).

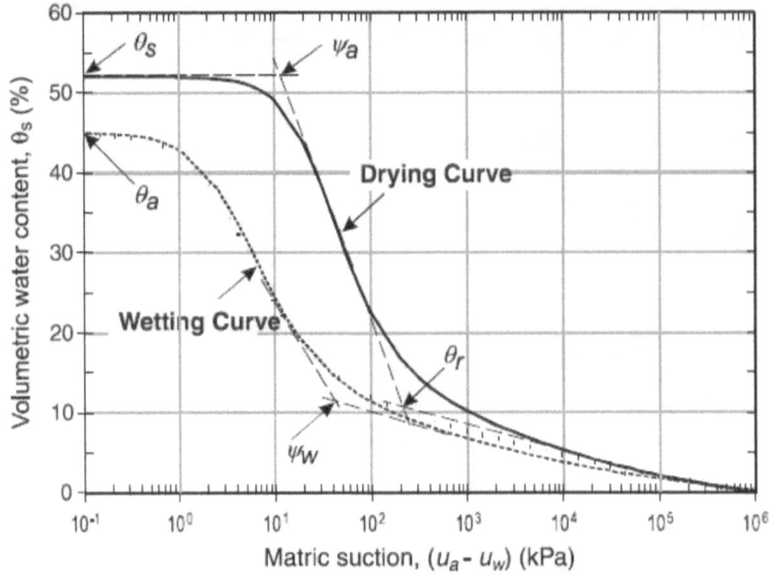

Figure 5.3 Definitions of terms for a typical soil-water characteristics curve (Tami et al. 2004)

The hydraulic conductivity functions are given in Figure 5.4. There was a disagreement between the saturated hydraulic conductivity, k_s, value provided in Table 5.1 for the pea gravel in the paper by Tami et al. (2004) and k_s value shown in Figure 5.4 in their paper. Tami et al. (2004) stated that the hydraulic properties of both the pea gravel and the concrete sand were based on the experimental work of Stormont and Anderson (1999). In the present research, k_s value for the pea gravel is taken from the paper by Stormont and Anderson (1999).

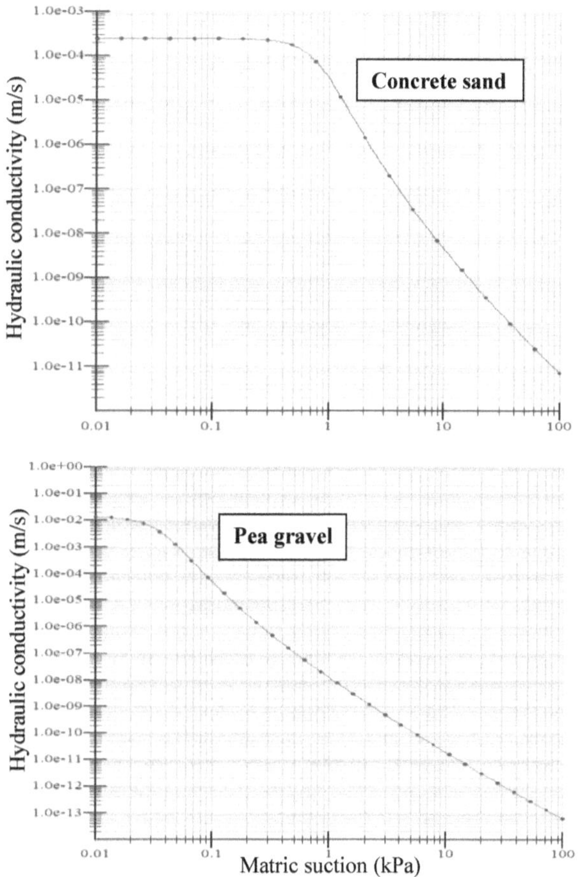

Figure 5.4 Hydraulic conductivity functions for two different soils

5.3 Numerical model in the present study for the physical model of sloping capillary barrier (as published in Tami et al. 2004)

5.3.1 Input information

The geometry of the domain analyzed in this numerical simulation is shown in Figure 5.5.

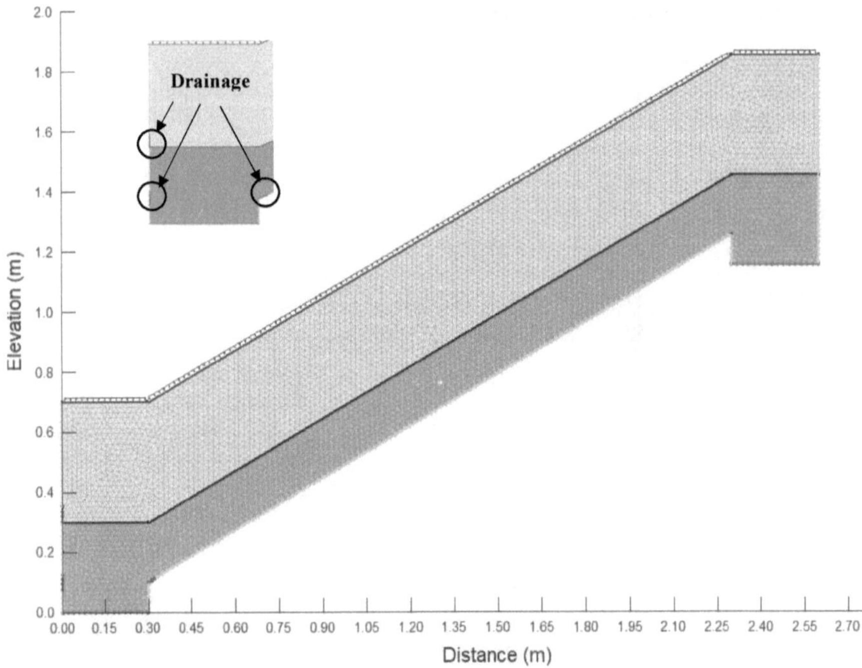

Figure 5.5 Geometry of the finite element domain in the present numerical analysis for the physical model of sloping capillary barrier reported by Tami et al. (2004)

The two-layer capillary barrier system was constructed using concrete sand (yellow region in the above figure) as the fine grained soil layer and pea gravel (brown region) as the coarse grained soil layer. The pressure head distribution within the system depends on many design factors namely, the thickness of soil layers, type of soil in each layer, precipitation rate, slope angle, and the hydraulic and thermal properties of the materials.

The dimensions of the laboratory model were 2.60 m in horizontal length, 0.4 m in width, and 0.70 m in height at the crest and toe of the model, as shown in Figure 5.5. The model was constructed using a slope angle of 30°.

A zero water flux boundary condition was applied at the left and right side of the model and at the bottom of the model. The water table (zero pressure boundary) was assumed at the bottom of crest and toe of the model. The precipitation boundary condition was applied/exposed at the top surface of the model. Thus, only the top surface will be affected by the precipitation event during the analysis. The precipitation was simulated as a positive flux boundary. A precipitation rate of 86.4 mm/h was applied for a duration of 42 min.

Drainage was also considered in the laboratory model and in the simulations by both Tami et al. (2004) and the present study. Drainage was included in the analysis for two reasons, namely for breakthrough collection (one outlet is located at the bottom of the fine soil layer, and the other one is located at the bottom of the coarse grained soil layer) as shown in Figure 5.5, and for lateral drainage collection (outlet is located at the bottom boundary of the model). The drainage boundaries were simulated as a unit hydraulic gradient boundary. Tami et al. (2004) explained the physical meaning of this boundary condition as the amount of water flux that passes through the boundary (i.e., the drainage boundary) at a particular matric suction and is equal to the hydraulic conductivity of the soil corresponding to that matric suction.

The water table was considered in the simulation as a zero pressure head boundary condition, and as a result, the total hydraulic head becomes equal to the elevation head. Table 5.2 shows a summary of the boundary conditions used in the analysis.

Table 5.2 Summary of the boundary conditions used in the present numerical analysis for the physical model of the sloping capillary barrier reported by Tami et al. 2004

Color	Name	Category	Kind	Parameters
	Drainage Outlet	Hydraulic	Water Unit Gradient	
	Positive water flux (86.4mm/hr)	Hydraulic	Water Flux	Infiltration
	Zero Pressure head	Hydraulic	Water Pressure Head	0 m
	Zero Water flux	Hydraulic	Water Flux	0 m/hr

5.3.2 Results of the present numerical analysis for the physical model of sloping capillary barrier (reported by Tami et al. 2004)

Before any comparisons are made between the results of the present study and the results of the numerical analysis of Tami et al. (2004), a series of figures (Figure 5.6 to Figure 5.16) are presented to illustrate how the infiltration progresses through the soil layers of the capillary barrier subjected to precipitation of limited duration. The simulation in the present study starts with the initial values equal to the values used as input by Tami et al. (2004).

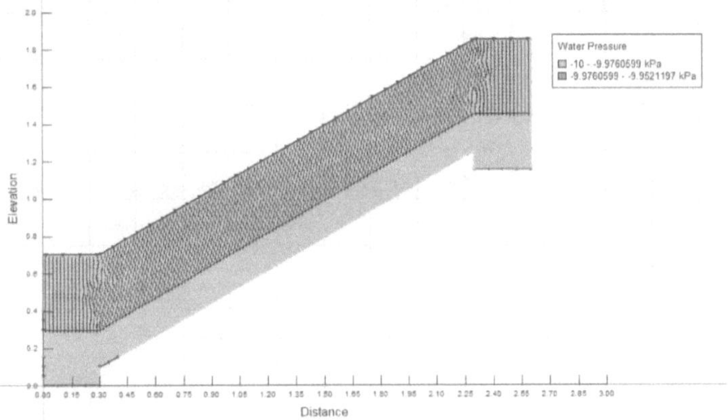

Figure 5.6 Initial pore water pressure (identical to the values used as input by Tami et al. 2004)

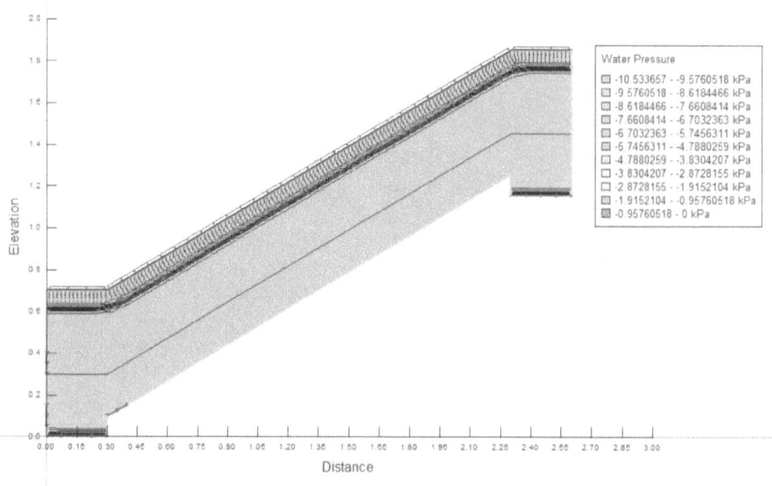

Figure 5.7 Calculated pore water pressures 10 minutes after the initiation of infiltration (present numerical analysis)

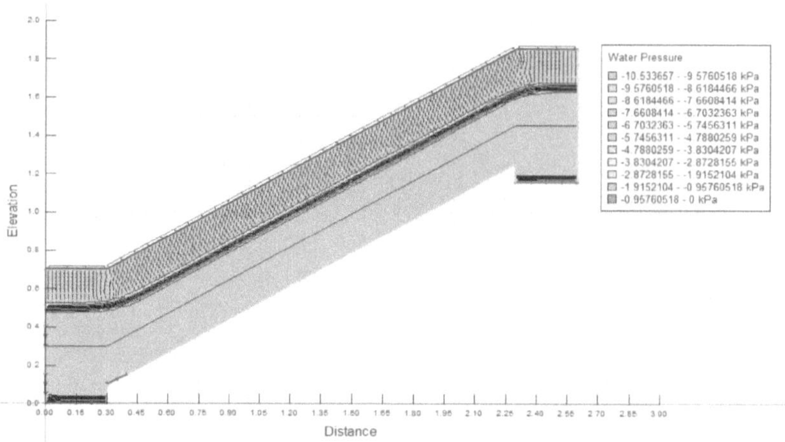

Figure 5.8 Calculated pore water pressures 20 minutes after the initiation of infiltration (present numerical analysis)

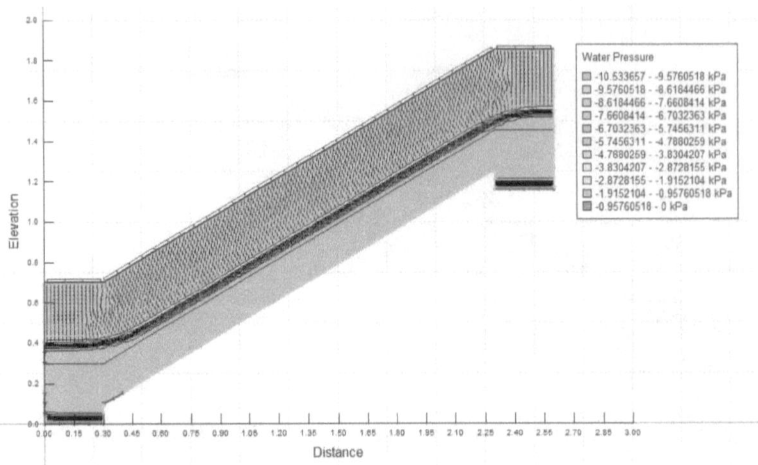

Figure 5.9 Calculated pore water pressures 30 minutes after the initiation of infiltration (present numerical analysis)

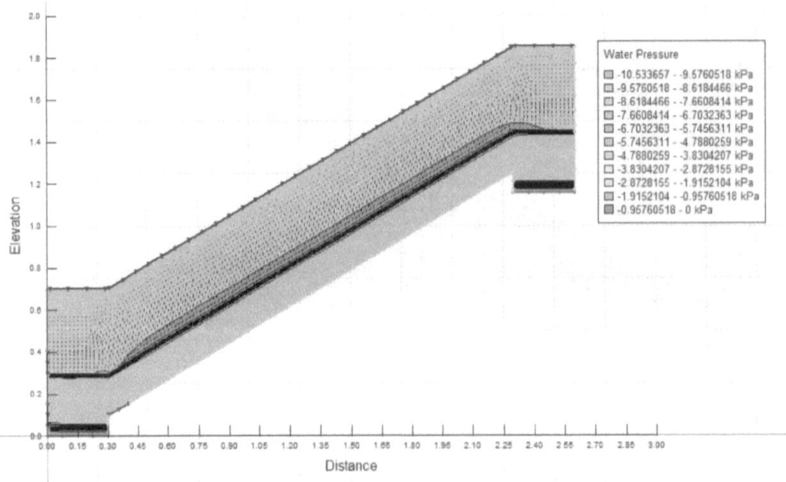

Figure 5.10 Calculated pore water pressures 40 minutes after the initiation of infiltration (present numerical analysis)

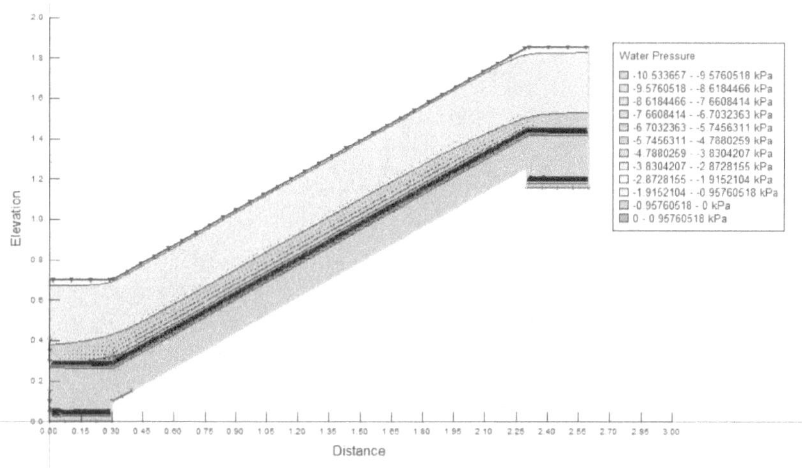

Figure 5.11 Calculated pore water pressures 50 minutes after the initiation of infiltration (present numerical analysis)

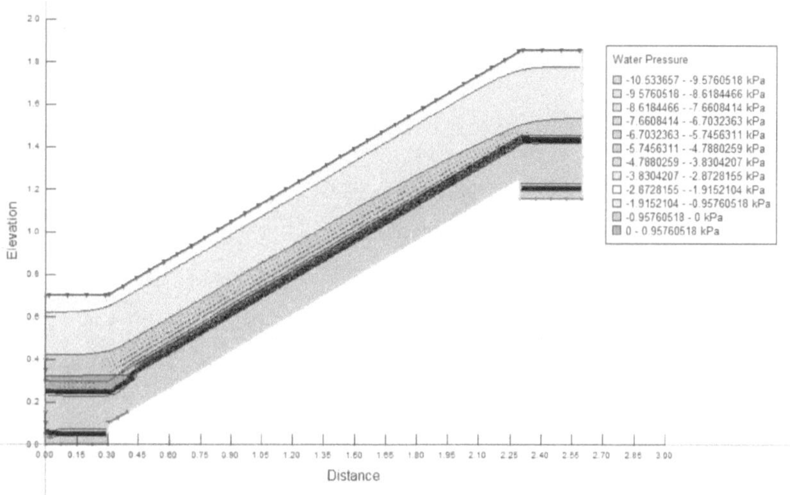

Figure 5.12 Calculated pore water pressures 60 minutes after the initiation of infiltration (present numerical analysis)

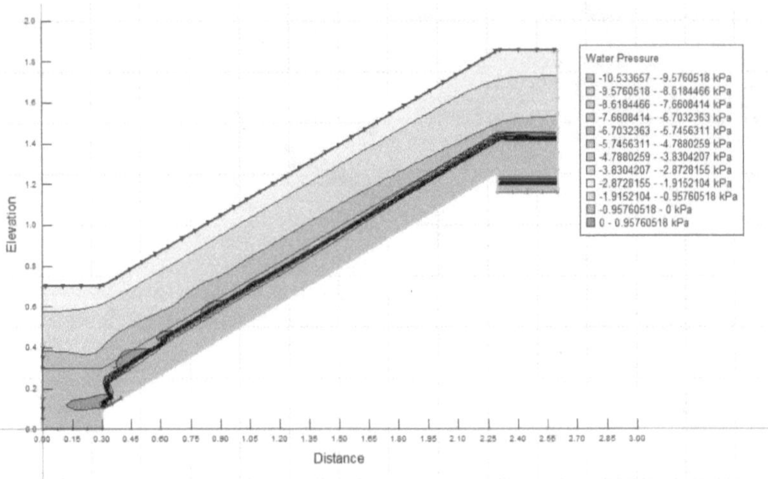

Figure 5.13 Calculated pore water pressures 70 minutes after the initiation of infiltration (present numerical analysis)

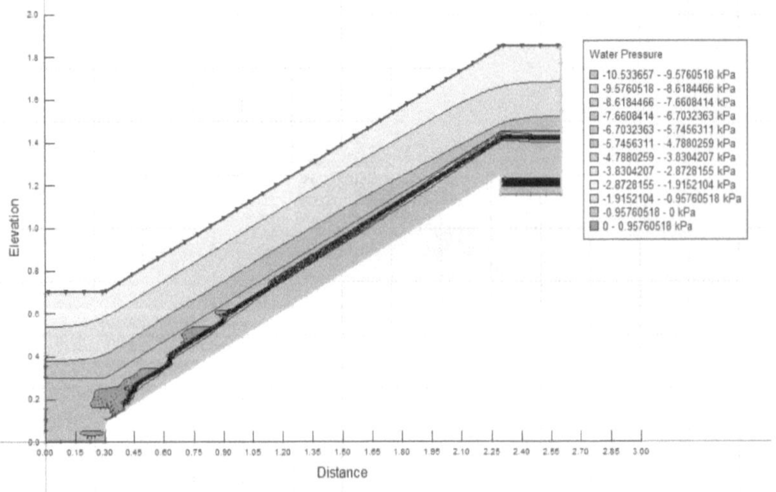

Figure 5.14 Calculated pore water pressures 80 minutes after the initiation of infiltration (present numerical analysis)

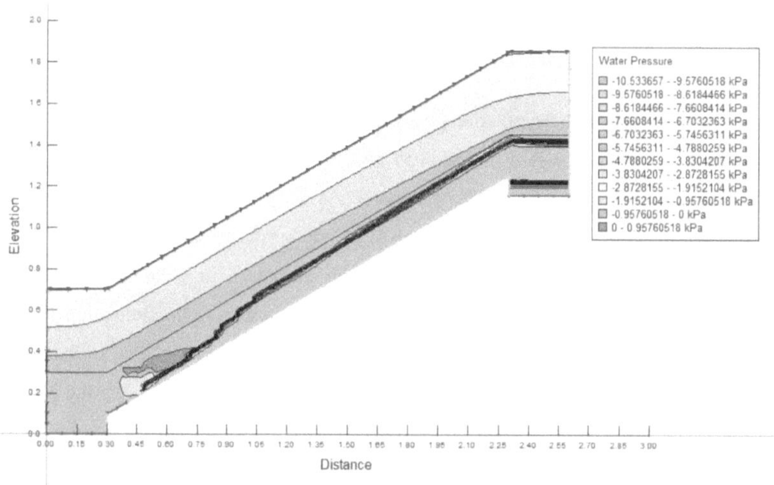

Figure 5.15 Calculated pore water pressures 90 minutes after the initiation of infiltration (present numerical analysis)

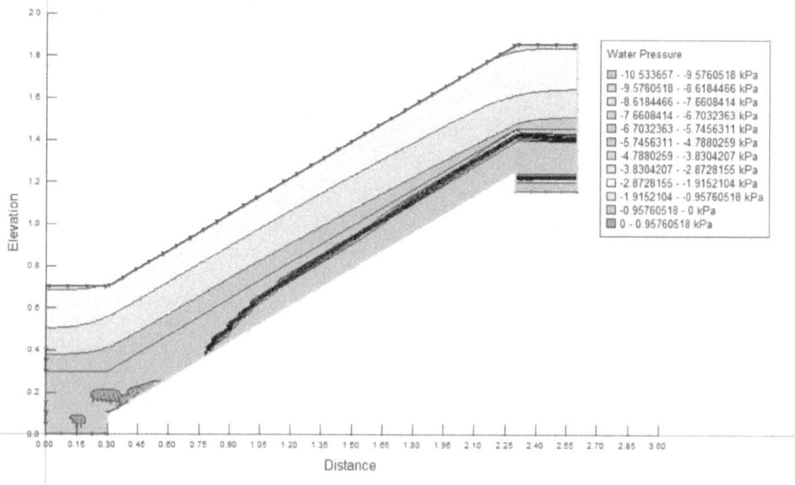

Figure 5.16 Calculated pore water pressures 100 minutes after the initiation of infiltration (present numerical analysis)

In the remaining part of this section, comparisons made between the results of the present study and the results of the numerical analysis of Tami et al. (2004) are provided.

In Figure 5.17, the results of simulations by Tami et al. (2004) and the results of simulations made in the present study are shown. The capillary barrier consisted of a fine-grained soil layer (concrete sand) placed over a layer of pea gravel. As shown in the figure, the input for the infiltration into the concrete sand at the surface of the slope was 10% of the saturated hydraulic conductivity (k_s) of the concrete sand. The infiltration continued for a duration of 42 minutes. There was no more precipitation or infiltration after 42 minutes. The simulation results are shown for three different vertical cross sections of the slope located at $x = 0.25$, 1.00, and 1.75 m from the toe of the slope (x refers to the distance on the horizontal axis). During the infiltration, the pressure head at the surface of the slope increased from an initial value of -0.54 m to -0.1 m in both simulations. The pressure head in the middle of the concrete sand started to increase after 20 min of precipitation. Along the interface between the concrete sand and pea gravel layers, the pressure head started to increase after about 35 min of simulation. The increase in the pressure head occurred rapidly and reached 0.0 m after 47 min of simulation in the present study. Apart from some minor differences, the results of the present study and the results of the numerical analysis of Tami et al. (2004) were very close to each other.

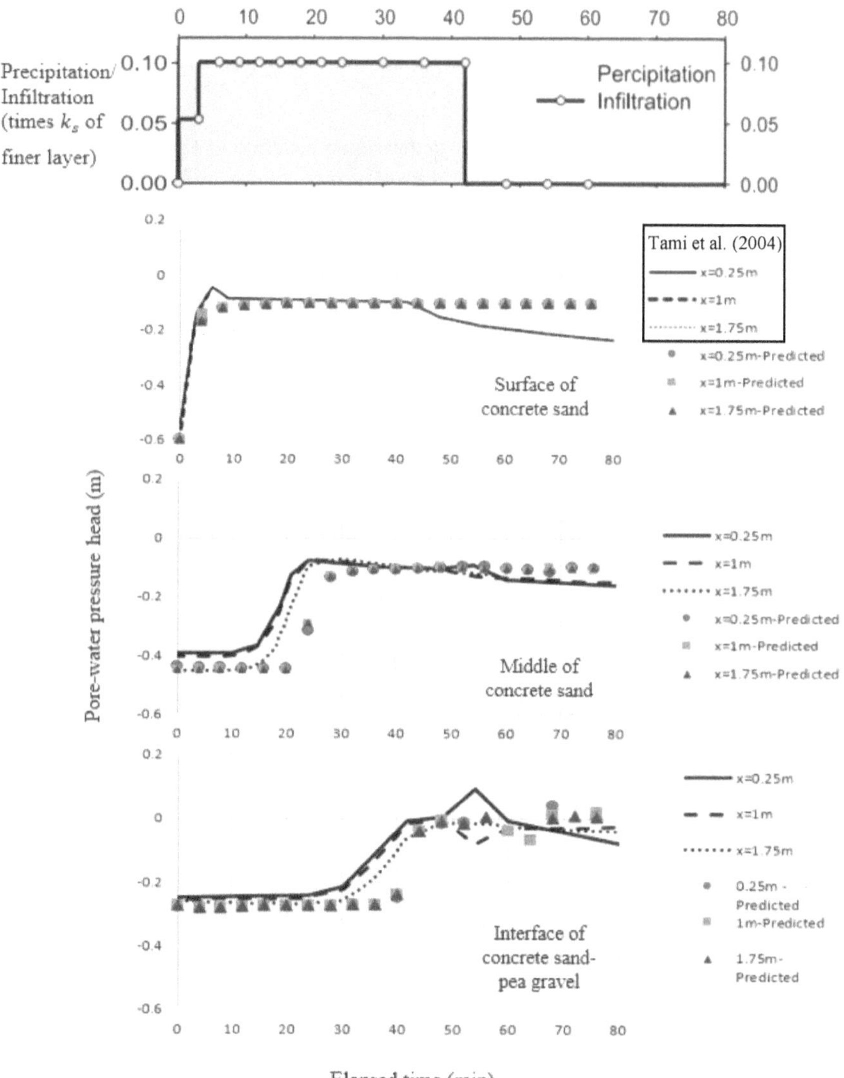

Figure 5.17 Variation of pressure head at various depths

Based on the comparison provided in Figure 5.17, it can be concluded that SEEP/W gives relatively good predictions for the behaviour of a sloping capillary barrier and that validates its use in the present study.

5.4 Numerical model in the present study to determine the effect of heated air flow on the behaviour of the physical model of sloping capillary barrier (reported by Tami et al. 2004)

The analysis described in this section makes use of SEEP/W and TEMP/W simultaneously to account for the coupled heat and fluid flow processes taking place in the model scale sloping capillary barrier, as described by Tami et al. (2004).

5.4.1 Input information

Figure 5.18 shows the geometry of the numerical model used in this section. It is identical to the model shown in Figure 5.5 of the previous section, except for the five perforated pipes that are added into the analysis domain. The pipes used in this section allow heated air to flow through them. Increasing temperature causes matric suction to increase particularly in the soil mass near the pipes. Thus, the maximum suction is expected to be generated in the soil element immediately next to a pipe. The value of the maximum suction can be calculated using Kelvin's equation. In this section, it is assumed that the temperature of the heated air reaches 40°C. All other boundary conditions are the same as explained in Section 5.3. Table 5.3 shows a summary of the boundary conditions used in the finite element analysis.

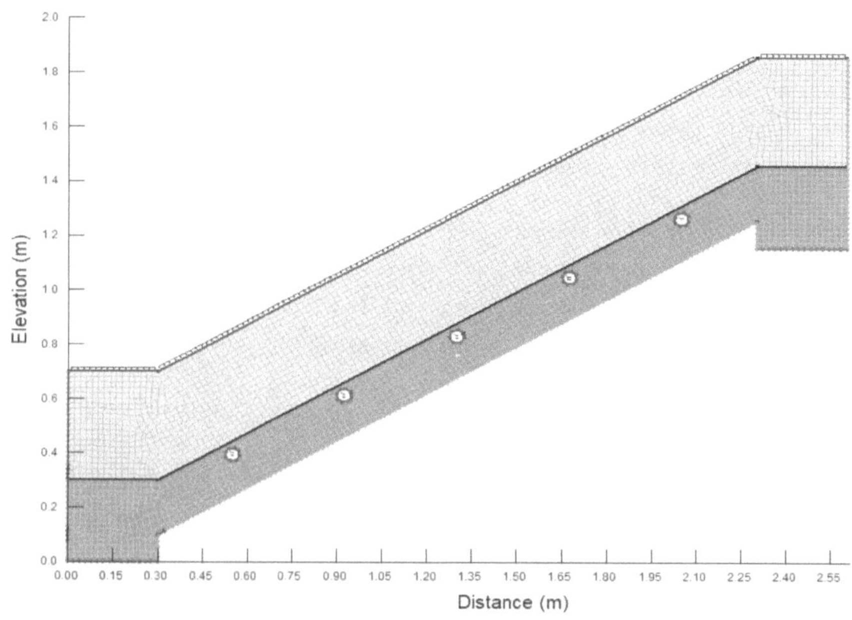

Figure 5.18 The geometry of the finite element model for the simulation of the behaviour of the capillary barrier subjected to heated air flow

Table 5.3 Boundary conditions for the finite element model with heated air flow

Color	Name	Category	Kind	Parameters
	Control Suction	Hydraulic	Water Pressure Head	-31,617.722 m
	Drainage Outlet	Hydraulic	Water Unit Gradient	
	Positive water flux (86.4mm/hr)	Hydraulic	Water Flux	Infiltration
	Temperature	Thermal	Temperature	40 °C
	Zero Pressure head	Hydraulic	Water Pressure Head	0 m
	Zero Water flux	Hydraulic	Water Flux	0 m/hr

Initial conditions for the finite element model

The initial pore water pressure head at the beginning of the transient analysis was specified as -1m which is equivalent to a matric suction value of -10 kPa. Same as in Section 5.3, this value was selected to simulate the dry condition of the capillary barrier system. The initial temperature was entered as 20°C. The maximum temperature of the heated air was selected to be 40°C.

Hydraulic properties of the soils

The volumetric water content as a function of matric suction (SWCC) for the concrete sand and pea gravel were generated using the Fredlund and Xing's model (Fredlund and Xing 1994). Table 5.4 shows the hydraulic parameters used in the analysis.

Table 5.4 Input data for the hydraulic properties used in the numerical analysis (Tami et al. 2004)

Material	Parameter							
	k_s(m/s)	θ_s	Ψ_a(kPa)	θ_r	Ψ_w(kPa)	α(kPa)	n	m
Concrete sand	2.4×10^{-4}	0.35	0.50	0.06	1.30	0.64	3.33	0.80
Pea gravel	1.3×10^{-2}	0.33	1.10	0.05	0.05	0.03	3.04	1.03

The curves for the unsaturated hydraulic conductivity versus matric suction functions for the concrete sand and pea gravel were generated by SEEP/W software, as shown in Figure 5.19.

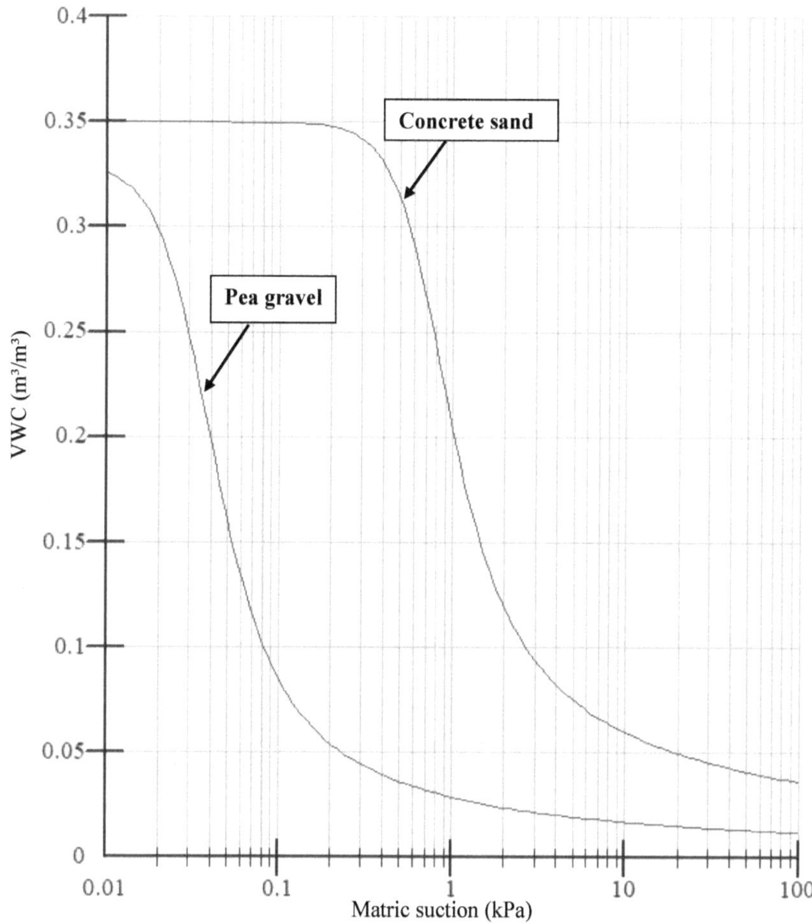

Figure 5.19 Volumetric water content versus matric suction curves (SWCC) used in the numerical analysis

Figure 5.20 shows the hydraulic conductivity curves for the concrete sand and pea gravel as a functions of matric suction.

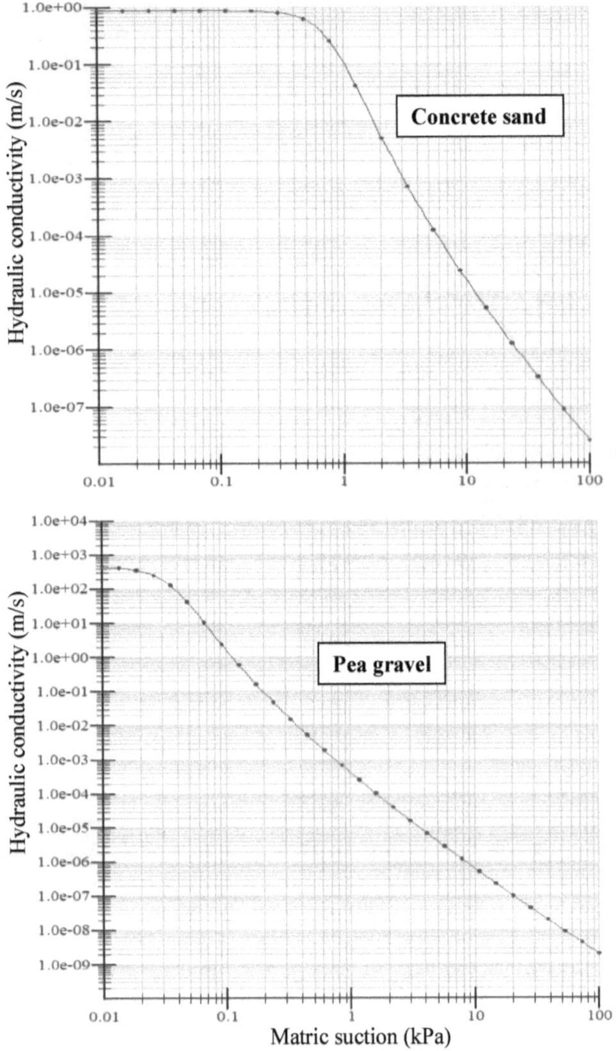

Figure 5.20 Hydraulic conductivity curves for the concrete sand and pea gravel

Thermal properties of the soils

Because the finite element model in this section was developed to study the effect of the heated air flow on the performance of the capillary barrier system, the material properties related to heat transfer analysis are required in addition to the hydraulic properties.

Figures 5.21 and 5.22 show the thermal conductivity versus volumetric water content for concrete sand and pea gravel, respectively.

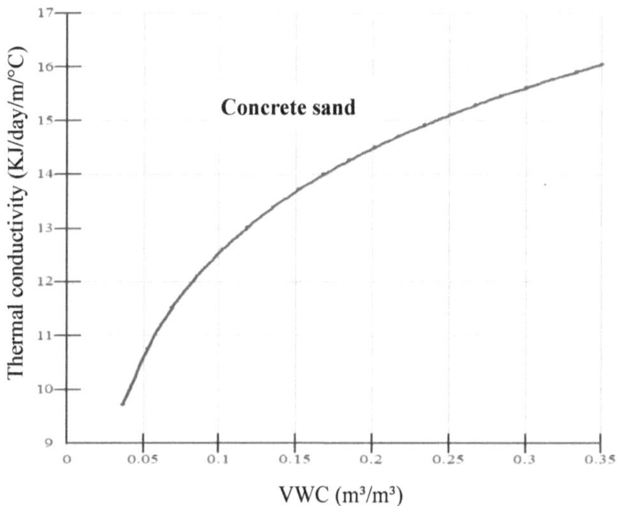

Figure 5.21 Thermal conductivity vs. VWC function for concrete sand

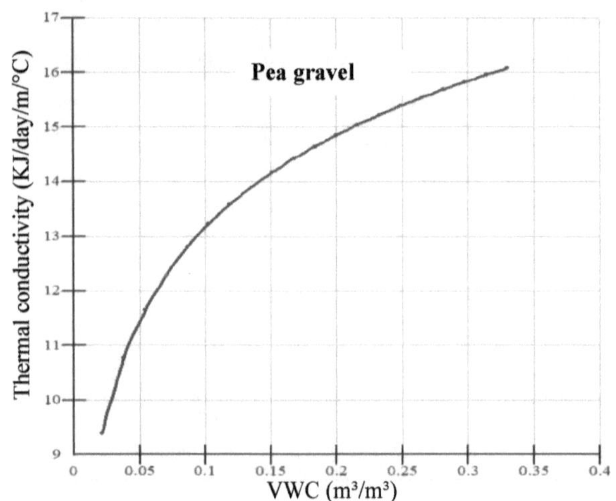

Figure 5.22 Thermal conductivity vs. VWC function for pea gravel

Figures 5.23 and 5.24 show the volumetric specific heat capacity functions for concrete sand and pea gravel, respectively.

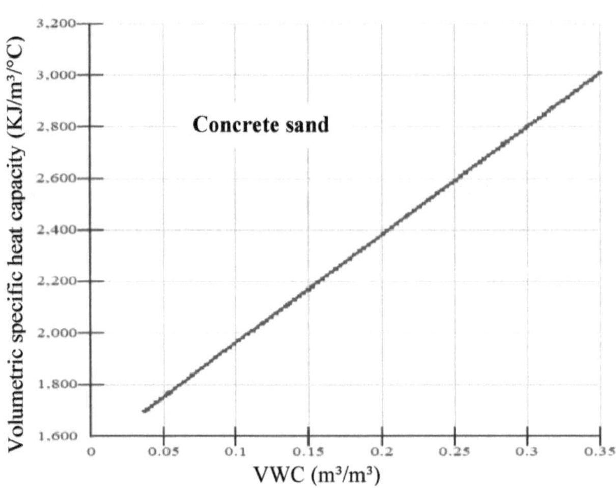

Figure 5.23 Volumetric specific heat capacity function for concrete sand

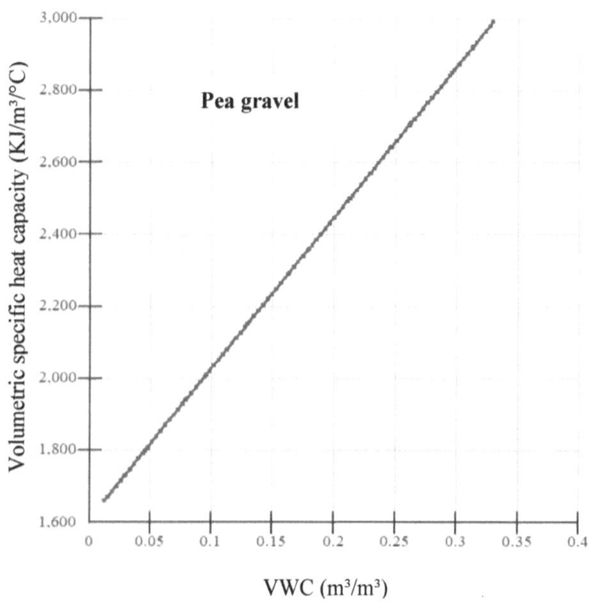

Figure 5.24 Volumetric specific heat capacity function for pea gravel

5.4.2 Results of the numerical analysis utilized the present study to determine the effect of heated air flow on the behaviour of the physical model of sloping capillary barrier (reported by Tami et al. 2004)

The temperature distribution in the soil layers resulting from the heated air flow in the five pipes is shown in Figure 5.25. As seen in the figure, the temperature is increasing everywhere in the layer of pea gravel and in about one half of the concrete sand. The temperature at the pipe perimeter is 40°C and its value decreases at points away from the pipes.

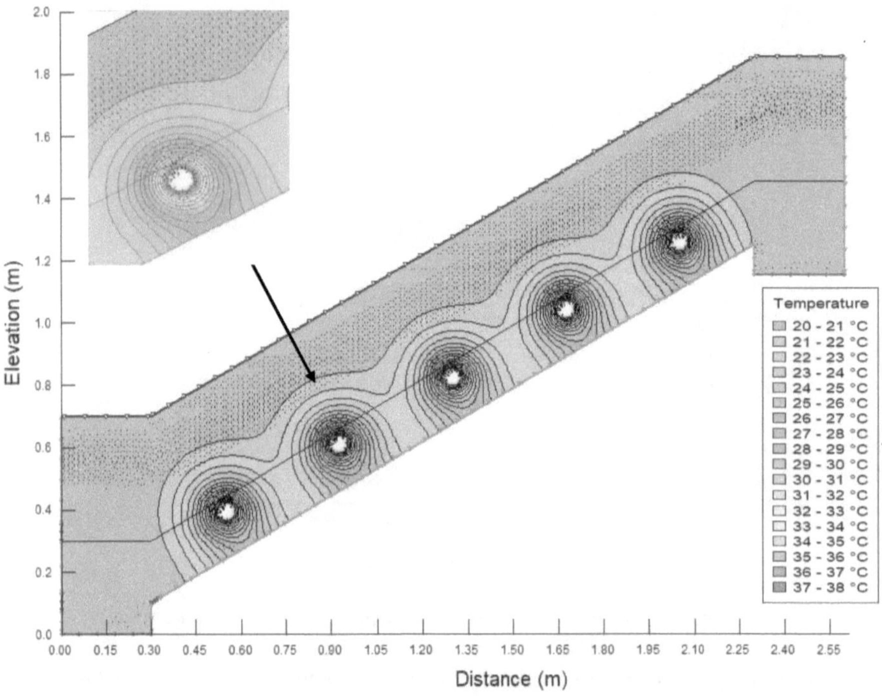

Figure 5.25 Temperature changes in the soil layers resulting from heated air flow

Figure 5.26 shows the velocity vectors of fluid flow around a pipe. The figure indicates that not only the VWC of the pea gravel decreases around the pipe, but the water in the concrete sand is also drained into the pipe.

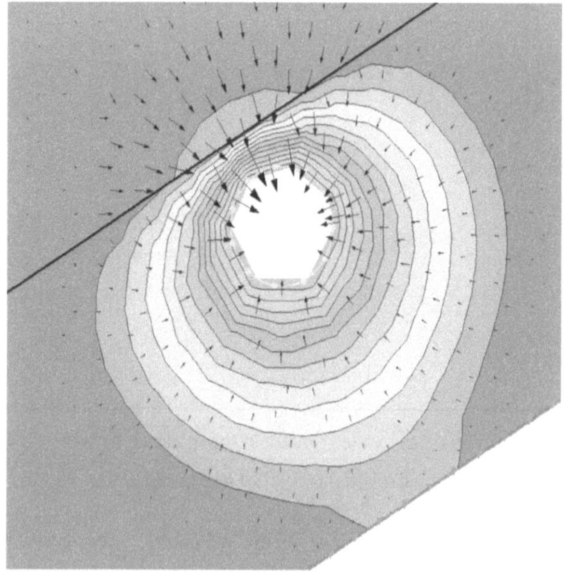

Figure 5.26 Velocity vectors of fluid flow toward a pipe

Volumetric water content changes over time

Figure 5.27 shows the variations in the volumetric water contents at three points located at the intersections between the concrete sand/pea gravel interface and the vertical lines at $x = 0.15$ m, 1m, and 1.75 m. Due to the heating effect from the heated air flow in the pipes, the VWC values at all three points decreased from a value of 0.038 (m^3/m^3) to 0.01 (m^3/m^3). This is a very significant change showing the benefits of heated air flow.

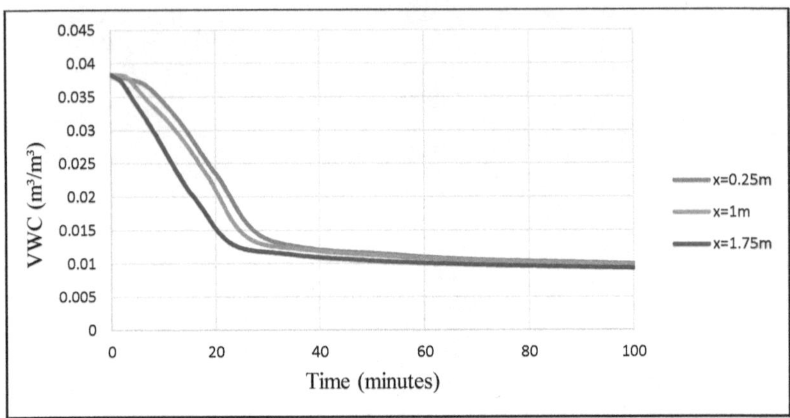

Figure 5.27 Variation of volumetric water content at the interface between the layer of concrete sand and the layer of pea gravel

Similarly, Figure 5.28 shows the variations in the volumetric water contents at three points located along the vertical lines at $x = 0.15$ m, 1 m, and 1.75 m in the middle of the pea gravel. Due to the effect of heating, the VWC values these three points decreased from a value of 0.017 (m^3/m^3) to about 0.003 (m^3/m^3). This is also a significant change showing the benefits of heated air flow.

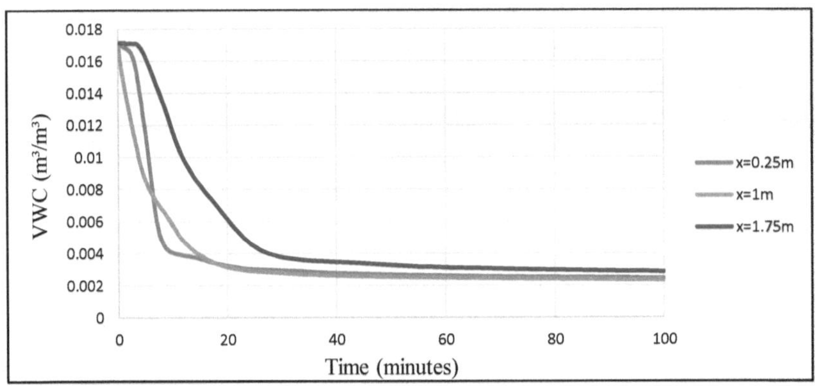

Figure 5.28 Variation of volumetric water content in the middle of pea gravel

Pore water pressure head changes over time

Figure 5.29 shows the changes in the pore water pressure heads at three points located at the intersections between the concrete sand/pea gravel interface and the vertical lines at $x = 0.15$ m, 1m, and 1.75 m. Due to the effects of heating, and the resulting decrease in the VWC, the pore water pressures heads at these three points decreased from -1 m to a range of values between -2400 m and -3000 m. In terms of matric suction, the increase was from 10 kPa to about 24000 kPa and 30000 kPa. This is a very large change in the matric suction showing the benefits of the heated air flow.

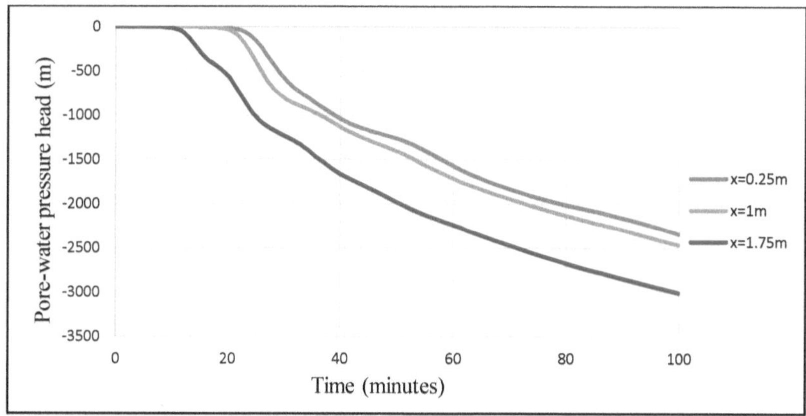

Figure 5.29 Variation of pore water pressure at the interface between concrete sand and pea gravel

Figure 5.30 shows the changes in the pore water pressure heads at three points located along the vertical lines at $x = 0.15$ m, 1 m, and 1.75 m in the middle of the pea gravel. Due to the effects of heating and the decrease in the VWC, the pore water pressures heads at these three points decreased from -1 m to a range of values between -12000 m and -18000 m. In terms of matric suction, the increase is from 10kPa to about 120000 kPa and 180000 kPa. This is a very large change in the matric suction showing the benefits of the heated air flow.

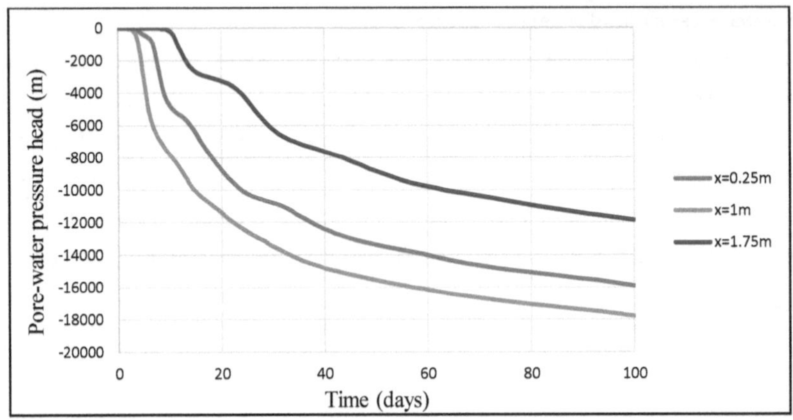

Figure 5.30 Variation of pore water pressure in the middle of pea gravel

5.5 Summary

The purpose of this chapter was to introduce heated air flow through the coarse-grained soil layer of the model scale tests of Tami et al. (2004) to determine its effect on the performance of this sloping capillary barrier. In order to achieve this objective, two numerical analyses were conducted. The first analysis was done to simulate the behaviour of the model scale capillary barrier which was conducted in room temperature. The simulations made in the present study were very close to the simulations by Tami et al. (2004). The favorable comparison of results suggests that the software is capable of producing acceptable predictions for the behaviour of capillary barriers and it is used correctly in the present study. In the second analysis, however, heated air flow was added into the numerical model. The comparisons of the results of this analysis and the results of the first analysis were done by plotting: (a) the volumetric water contents, and (b) matric suction values at different locations in the capillary barrier.

The numerical results presented in this chapter show that heated air flow has a significant influence on the performance of the capillary barrier system.

CHAPTER 6

CASE STUDY (2): AN INVESTIGATION TO DETERMINE THE EFFECT OF HEATED AIR FLOW IN THE PERFORMANCE OF SAINT-TITE-DES-CAPS MUNICIPAL LANDFILL COVER

6.1 Introduction

In this chapter, the second case study, CASE STUDY (2), is presented. This case study is based on the detailed information provided in a paper by Abdolahzadeh et al. (2011). The subject is a landfill cover with capillary barrier constructed over the municipal waste at Saint-Tite-des-Caps, Quebec, Canada. There are several other publications by the same group of authors (Vachon et al. 2015; Abdolahzadeh et al. 2008; Abdolahzadeh et al. 2011b) dealing with various aspects of the cover for this landfill site (i.e., Saint-Tite-des-Caps), such as the assessment of its design by comparing design parameters against field measurements, prediction of the diversion length, and the steady state and transient analyses of the hydraulic behaviour of the capillary barrier. The construction includes an instrumented section of the landfill cover, as shown in Figure 6.1. The top layer, labelled "Protection layer" in the figure, protects the lower layers and is required by Quebec landfill regulations (Vachon et al. 2015). The layer below the "Protection layer" is labelled DBP, which is 0.6 m thick and it stands for deinking by-product. DBP acts as a seepage control layer. The actual capillary barrier is located below the DBP. The capillary barrier consists of a 0.4 m thick layer of sand placed on a 0.2 m thick gravel layer.

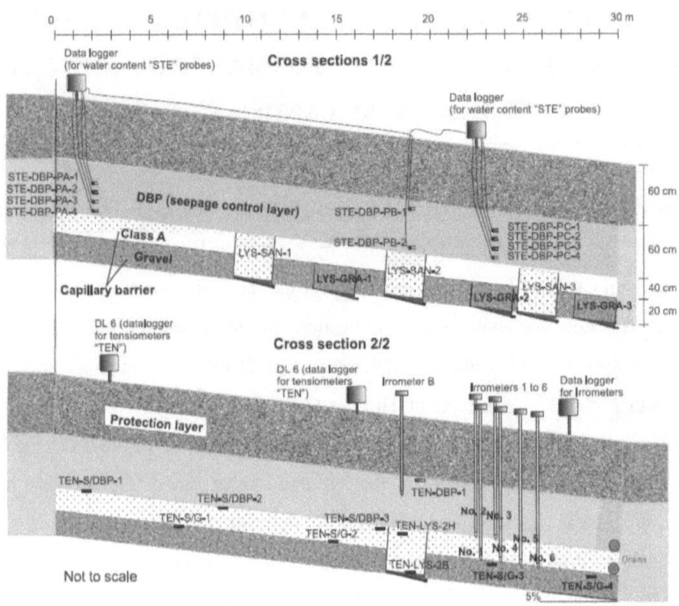

Figure 6.1 Two cross-sectional views of the Saint-Tite-des-Caps at the experimental part of the landfill cover (from Abdolahzadeh et al. (2011a)

Abdolahzadeh et al. (2011a), Abdolahzadeh et al. (2011b), and Vachon et al. (2015) provided detailed information on how the landfill cover was designed, all relative information for the layers of unsaturated/saturated soils, the initial conditions, boundary conditions, and the measured field behaviour recorded over a period of four years.

The purpose of this chapter is to introduce heated air flow through the coarse-grained soil (gravel) layer of the landfill cover in order to determine its effect on the performance of the capillary barrier. Two numerical analyses were conducted to achieve this objective. The first analysis, named FE-Analysis (No Heated Air flow), was done to simulate the behaviour of the capillary barrier without considering the effect of heated air flow. In the second analysis, named FE-Analysis (With Heated Air flow), heated air flow was added to the numerical model. The comparison of the results of these analyses was done by plotting: (a) the volumetric water contents, and (b) matric suction values.

6.2 FE Analysis (No Heated Air flow)

A two-dimensional FE analysis was conducted using SEEP/W. Detailed information about the analysis is provided in the following sections.

6.2.1 Geometry and the boundary conditions

Figure 6.2 shows the geometry used in the simulations. The thicknesses of the sand layer, gravel layer, and the waste layer were 0.4 m, 0.2 m and 0.5 m, respectively. The slope of 5% was used to construct the inclined capillary barrier system.

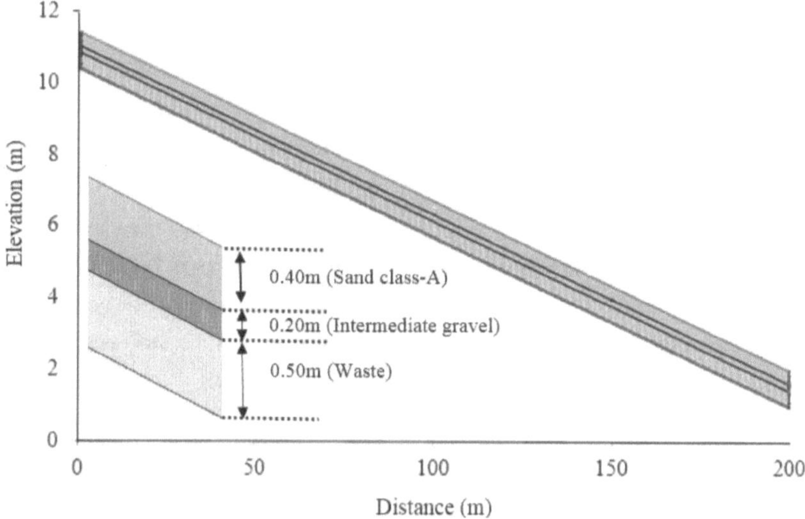

Figure 6.2 The geometry of the analysis domain for the Saint-Tite-des-Caps landfill cover

The hydraulic boundary conditions, used in the present study, were identical to the boundary conditions used in the FE analysis of Vachon et al. (2015) for the same landfill cover. Table 6.1 shows a summary of the hydraulic boundary conditions.

Table 6.1 Hydraulic boundary conditions

Color	Name	Category	Kind	Parameters
	Drainage	Hydraulic	Water Unit Gradient	
	Infiltration	Hydraulic	Water Flux	1e-08 m/sec
	Water Table	Hydraulic	Water Pressure Head	0 m
	Zero flux	Hydraulic	Water Flux	0 m/sec

Four types of boundary conditions were used in this analysis. The first one was a positive water flux boundary at the top of the sand layer. A uniform water flux of (1.0E-08 $m^3/sec/m^2$) was specified at the top of the capillary barrier system. This flux rate represents the maximum value that corresponds to the field observations. The second boundary condition was a unit hydraulic gradient boundary created at the lower edge of the cover to simulate the drainage (the physical meaning of this boundary was explained in CASE STUDY (1) in Chapter 5. The third was zero pressure boundary condition that was created at the bottom of the analysis domain to represent the water table. The fourth boundary condition was zero water flux that was specified for the vertical sections at the upstream end and downstream ends of the model.

6.2.2 Hydraulic properties of the waste and the soils used in different layers of the capillary barrier

Hydraulic properties of the municipal waste and the soils used in the numerical simulations are shown in Table 6.2.

Table 6.2 Hydraulic properties of the soils and the municipal waste used in the numerical simulations. (Vachon et al. 2015)

Parameter	Sand	Gravel	Waste
WRC model	vG	vG	vG
α (1/kPa)	0.472	1.953	0.38
n	6.32	4.20	1.47
m	0.842	0.762	0.32
θ_s (m³/m³)	0.33	0.35	0.30
θ_r (m³/m³)	0.05	0.07	0.01
k_s (m/s)	1.5×10^{-4}	1.5×10^{-3}	1.0×10^{-5}
ψ_{AEV} (kPa)	1.4	0.4	2.6
ψ_{WEV} (kPa)	3.5	1.7	200

The meaning of the parameters included in the table above are listed below:

WRC   : Water retention curve which is another term used for SWCC
vG   : van Genuchten
α : A parameter used in van Genuchten model for WRC
n : A parameter used in van Genuchten model for WRC
m : A parameter used in van Genuchten model for WRC
k_s : Saturated hydraulic conductivity
ψ_{AEV} : Suction at air-entry value
θ_s (m³/m³): Saturated water content
θ_r (m³/m³): Residual water content

The volumetric water content versus matric suction curves were generated by using the van Genuchten model (van Genuchten 1980). Figures 6.3, 6.4, and 6.5 show the SWCCs for the materials used in the landfill cover and in the municipal waste. Figures 6.6 through 6.8 show the hydraulic conductivity functions for the materials in the landfill cover and the landfill waste material.

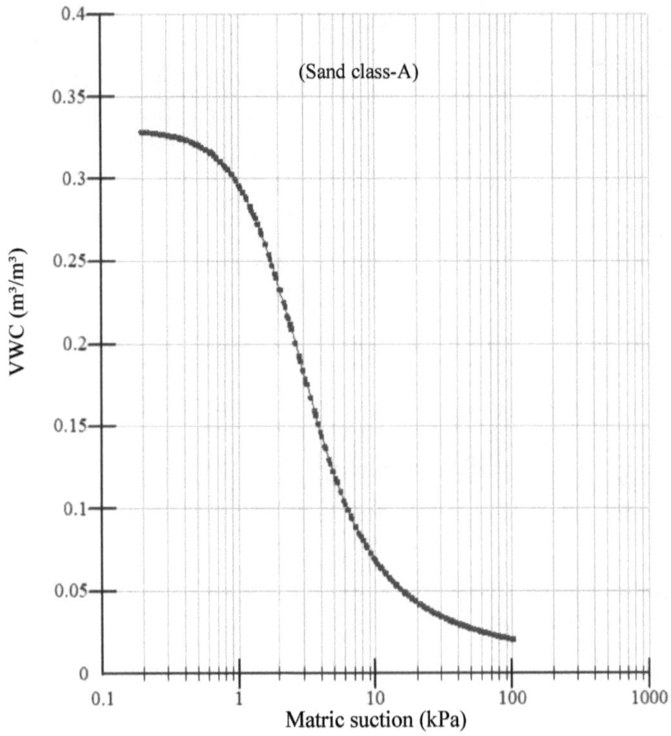

Figure 6.3 Volumetric water content versus matric suction curve (SWCC) of the sand used in the numerical analysis

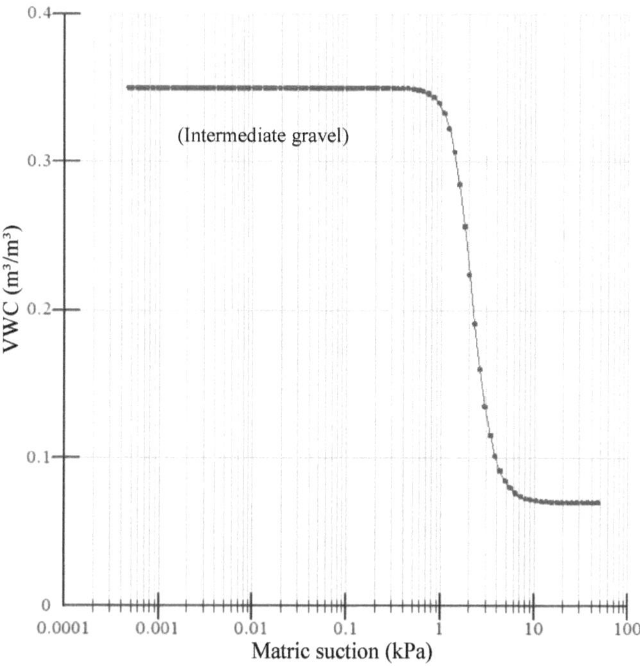

Figure 6.4 Volumetric water content versus matric suction curve (SWCC) of the gravel used in the numerical analysis

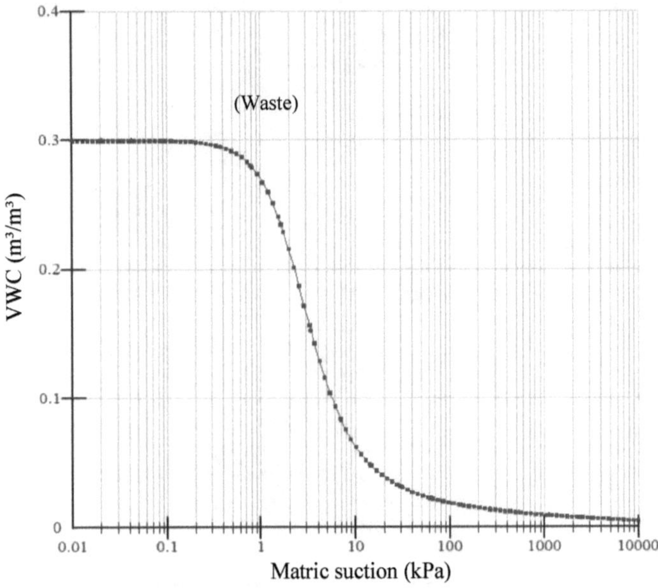

Figure 6.5 Volumetric water content versus matric suction curve (SWCC) of the municipal waste used in the numerical analysis

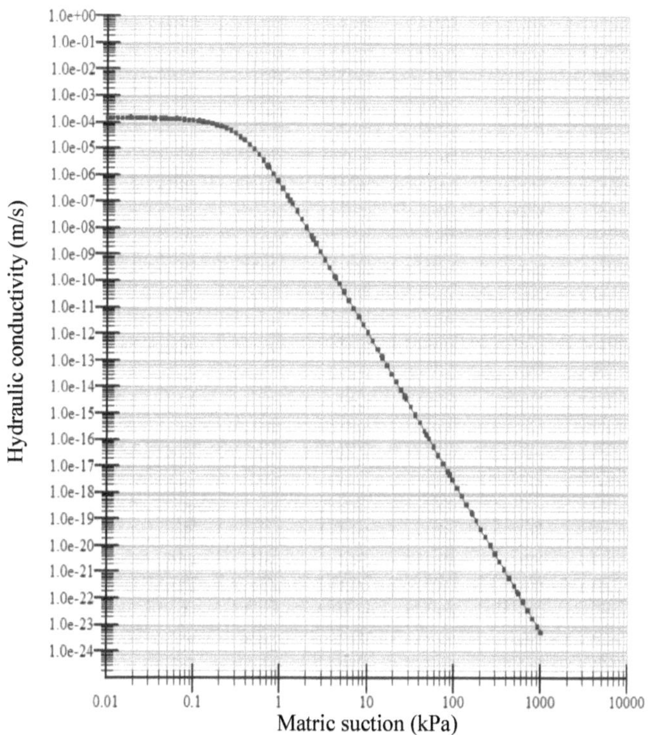

Figure 6.6 Hydraulic conductivity versus matric suction function for sand

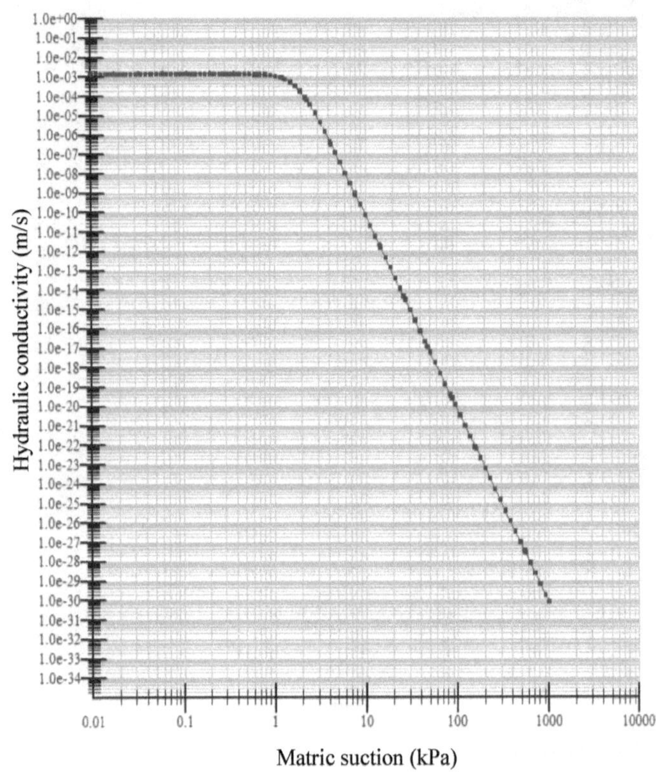

Figure 6.7 Hydraulic conductivity versus matric suction function for gravel

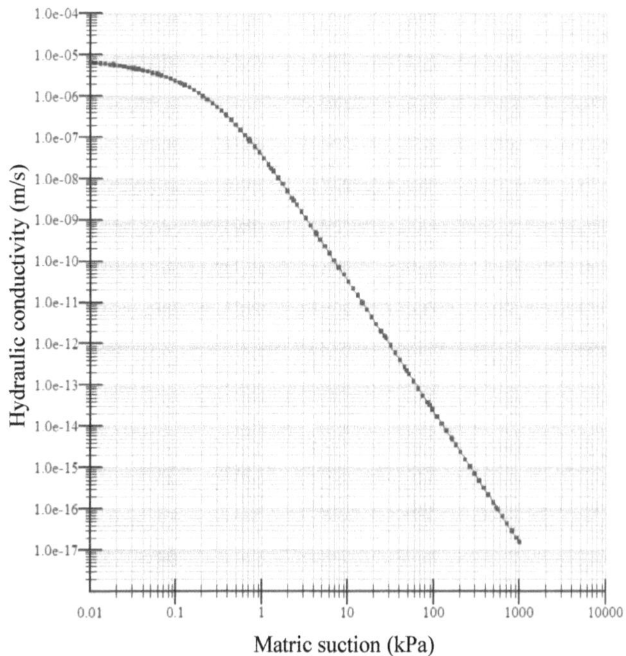

Figure 6.8 Hydraulic conductivity versus matric suction function for the municipal waste

6.2.3 Initial conditions

Similar to the initial conditions used in the analysis by Vachon et al. (2015), the initial conditions used in the present analysis were for the wet conditions (the worst-case scenario) for all layers of the cover. The activation pore-water pressures in the analysis were considered equal to the air-entry values for the landfill cover materials. The values of -1.4 kPa, -0.4 kPa, and -2.6 kPa were specified for sand, gravel, and waste materials, respectively.

6.2.4 Mesh size and time step

Different element size and mesh densities were used in the simulation to improve the solution accuracy at critical areas such as the sand-gravel interface. The thickness of the elements at the sand-gravel interface was 0.09 m. In contrast, the thickness of the elements in the municipal waste was 0.25 m. Smaller size of elements was used where there was a large gradient in pore water pressure to improve the solution accuracy for that particular region. The mesh at the sand-gravel interface was simulated using the "surface layer" option in the software. This type of element is beneficial to achieve convergence in case of rapid changes in boundary conditions (Heat and Mass Transfer Modeling with GEOSTUDIO, 2018).

The analysis was carried out for a total duration of 30 days. The time steps were selected on the basis of element size and hydraulic conductivity values.

6.2.5 Results of FE Analysis (No Heated Air flow)

In this section, the results of the finite element analysis without any heated air flow are presented in terms of the changes in VWC and matric suction.

6.2.5.1 Volumetric water content changes over time

Figures 6.9 and 6.10 show the variations in the volumetric water contents at six points. Three of these points were located at the intersections of the vertical lines at $x = 50$ m, 100 m, and 150 m and the line corresponding to the interface between the sand layer and gravel layer (Figure 6.9).

The other three points were chosen in the middle of the coarse grained soil layer on three vertical lines at $x = 50$ m, 100 m, and 150 m (Figure 6.10). The results showed that the VWC was higher at the points located at a larger distance from the top of the slope.

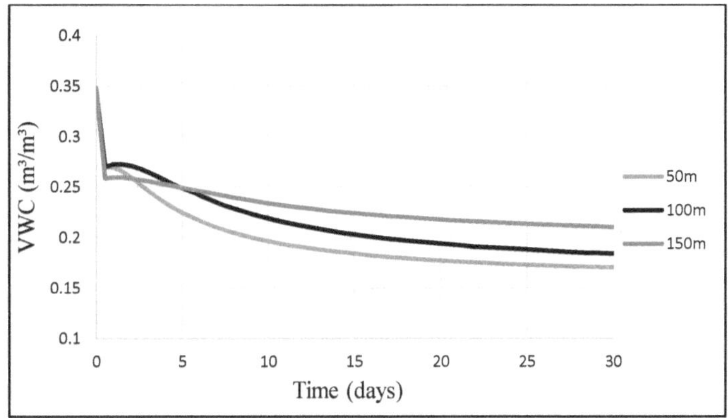

Figure 6.9 Variation of the volumetric water content at the sand-gravel interface

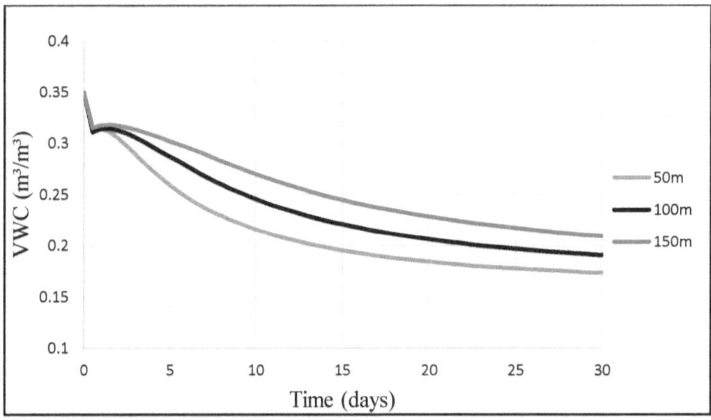

Figure 6.10 Variation of the volumetric water content in the middle of gravel layer

6.2.5.2 Pore water pressure changes over time

Figures 6.11 and 6.12 show the change in the pore water pressure at the same locations corresponding to the VWCs plotted in Figures 6.9 and 6.10. These results showed that, the pore water pressure (negative pore water pressure is equal to matric suction) was smaller at the points located at a larger distance from the top of the slope.

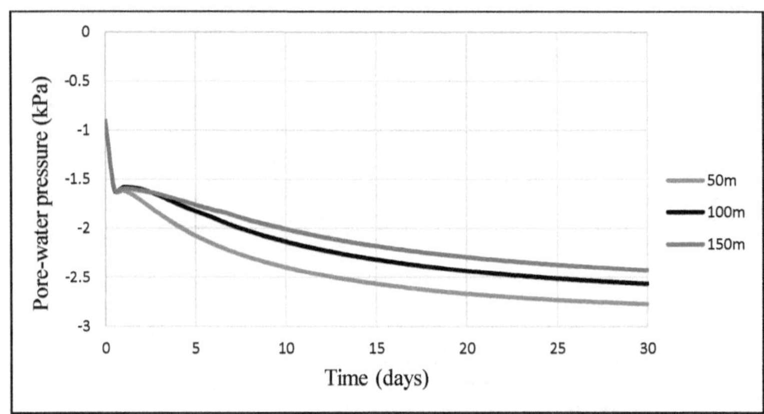

Figure 6.11 Variation of pore water pressure at the sand-gravel interface

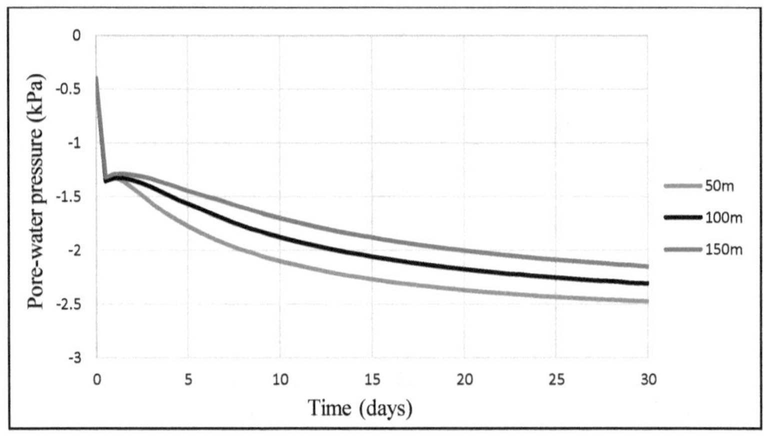

Figure 6.12 The variation of pore water pressure in the middle of gravel layer

6.3 FE Analysis (With Heated Air flow)

A transient FE analysis was conducted using SEEP/W and TEMP/W simultaneously to determine the effect of heated air flow on the cover performance. Details about the analysis are provided in the following sections.

6.3.1 Geometry and the boundary conditions

Pipes were added to the geometry used in the previous section to conduct the heated air flow. Figure 6.13 shows the geometry and the boundary conditions used in the FE Analysis (With Heated Air flow). The pipes were 5 cm in diameter and they were placed in the coarse-grained soil layer near the interface between the fine and the coarse soil layers.

In addition to the boundary conditions used in the FE analysis Without Heated Air flow (Section 6.2), thermal boundary conditions were needed in the analysis to include temperature effects on the behaviour of the capillary barrier. The temperature at the perimeter of the pipes was set at 35°C. The suction generated by the increased temperature was calculated to be 266524 kPa that was equivalent to a pressure head value of (-27,177 m). Kelvin's equation (Equation 2.50) was revisited again to calculate the suction due to temperature change.

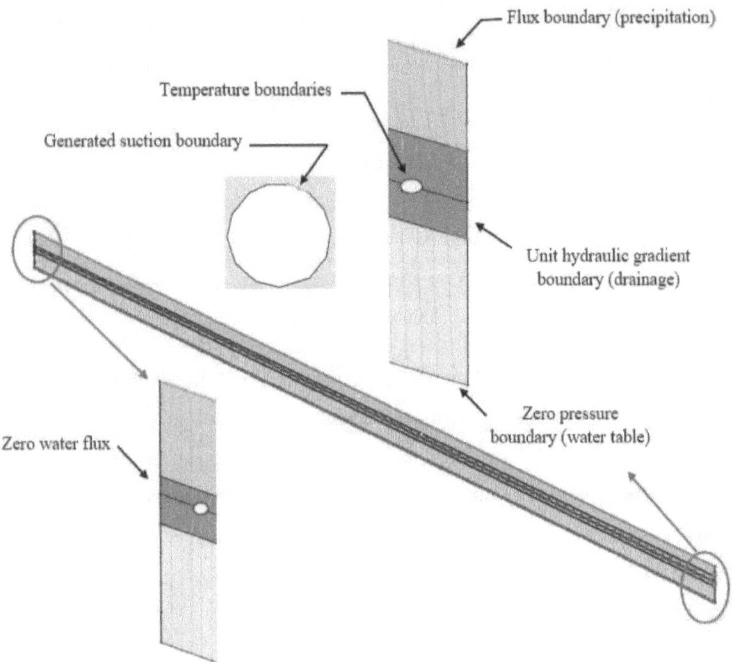

Figure 6.13 The geometry of Saint-Tite-des-Caps landfill final cover with the boundary conditions

A summary of the boundary conditions in this FE analysis with heated air flow is provided in Table 6.3.

Table 6.3 Hydraulic and thermal boundary conditions for model

Color	Name	Category	Kind	Parameters
	Control suction	Hydraulic	Water Pressure Head	-27,177 m
	Drainage	Hydraulic	Water Unit Gradient	
	Infiltration	Hydraulic	Water Flux	1e-08 m/sec
	Temperature	Thermal	Temperature	35 °C
	Water Table	Hydraulic	Water Pressure Head	0 m
	Zero flux	Hydraulic	Water Flux	0 m/sec

6.3.2 Hydraulic properties of the waste and the soils placed in different layers of the capillary barrier

The hydraulic properties of the municipal waste and the soils used in the numerical simulations were the same as in the FE analysis Without Heated Air flow (please see Table 6.2).

6.3.3 Thermal properties of the waste and the soils placed in different layers of the capillary barrier

The thermal properties of unsaturated soils are mainly a function of their volumetric water content. The coefficient of heat conduction and volumetric specific heat capacity are the required input parameters. The analysis started at 24°C for all layers of the capillary barrier. The initial values of these parameters before any changes in VWC due to infiltration are summarized in Table 6.4.

Table 6.4 Values of parameters for the thermal analysis

Function	The input parameters	Value			Unit
		Sand	Gravel	Waste	
Thermal conductivity vs. volumetric water content	Thermal conductivity	0.00311	0.00445	0.00284	kJ/sec/m/°C
Specific heat function	Volume specific Heat Capacity	1801.22	3009.5	1343.0632	kJ/m³/°C

Figures 6.14 through 6.16 show the thermal conductivity versus the volumetric water content functions for the materials included in the analysis domain. Figures 6.17 through 6.19 show the volumetric specific heat capacity for the same materials.

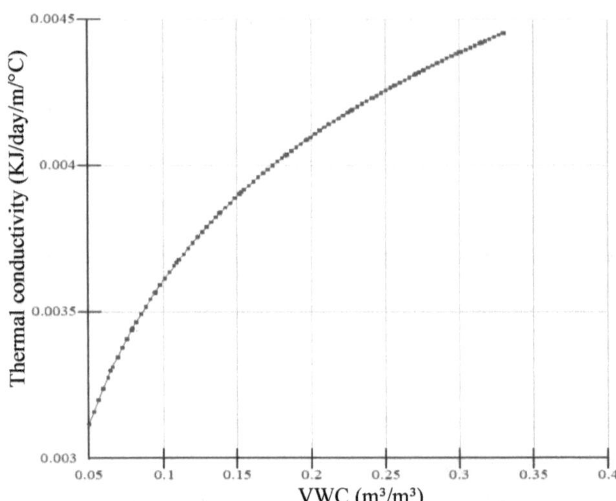

Figure 6.14 Thermal conductivity vs. VWC function for sand

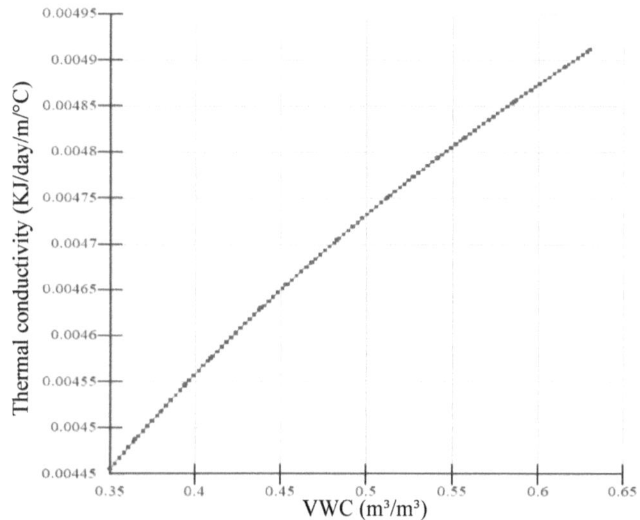

Figure 6.15 Thermal conductivity vs. VWC function for gravel

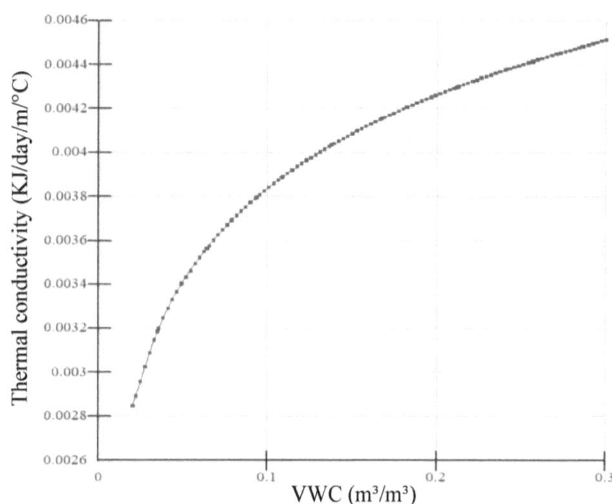

Figure 6.16 Thermal conductivity vs. VWC function for waste

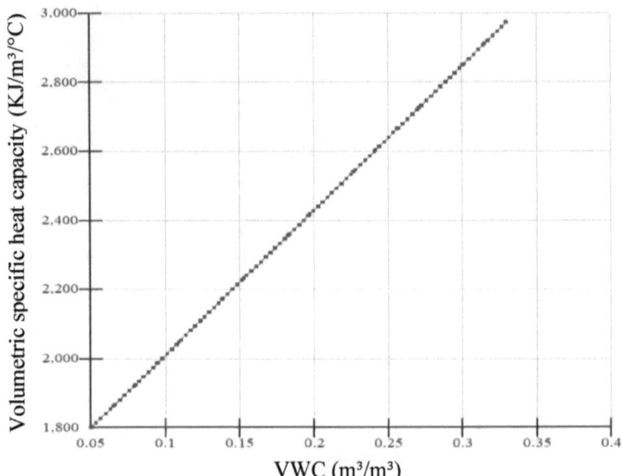

Figure 6.17 Volumetric specific heat capacity of sand

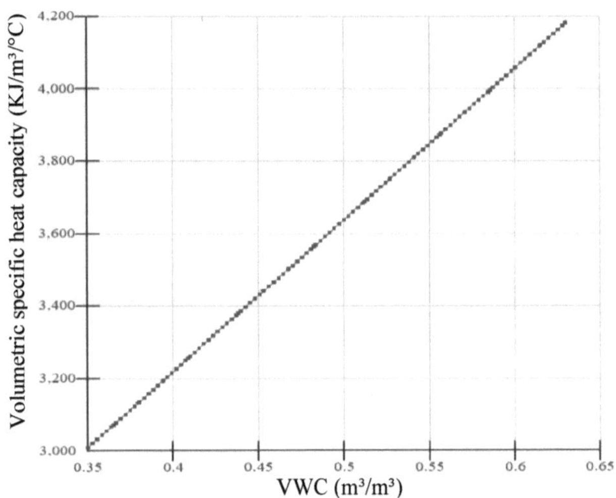

Figure 6.18 Volumetric specific heat capacity of gravel

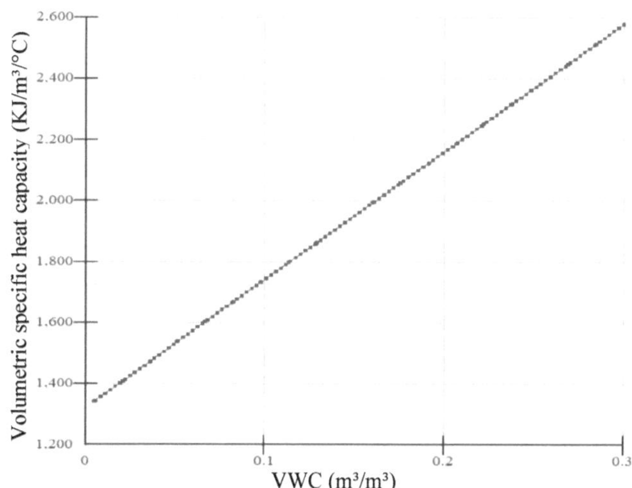

Figure 6.19 Volumetric specific heat capacity of municipal waste

6.4 Simulation results of the FE analysis with heated airflow

The results of the FE analysis with heated air flow are presented and discussed in terms of the changes in VWC and matric suction at the same six points as in the FE analysis without heated air flow.

6.4.1 Volumetric water content changes over time

Figure 6.20 and 6.21 show the variation in the volumetric water content during the analysis for the preheated air flow. Three observation can be made: (1) the distance along the slope measured from the top of the slope has very little influence on the changes taking place in VWC as a function of time; (2) heated air flow cause rapid reduction in the VWC as compared to the Figures 6.9 and 6.10 where there was no heated air flow; and (3) the amount of reduction in VWC is much larger in case of heated air flow.

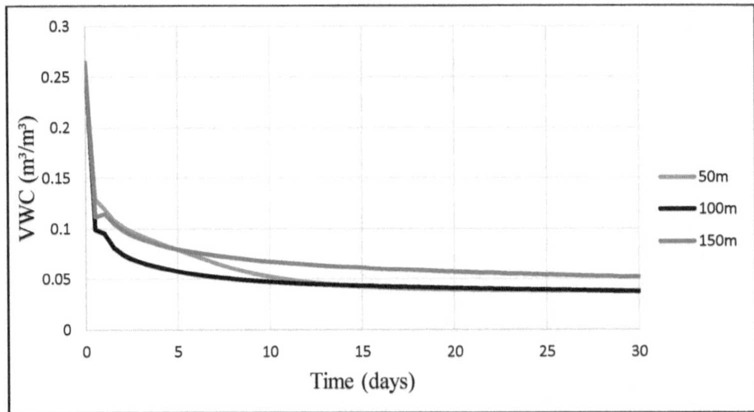

Figure 6.20 The variation of the volumetric water content at sand-gravel interface

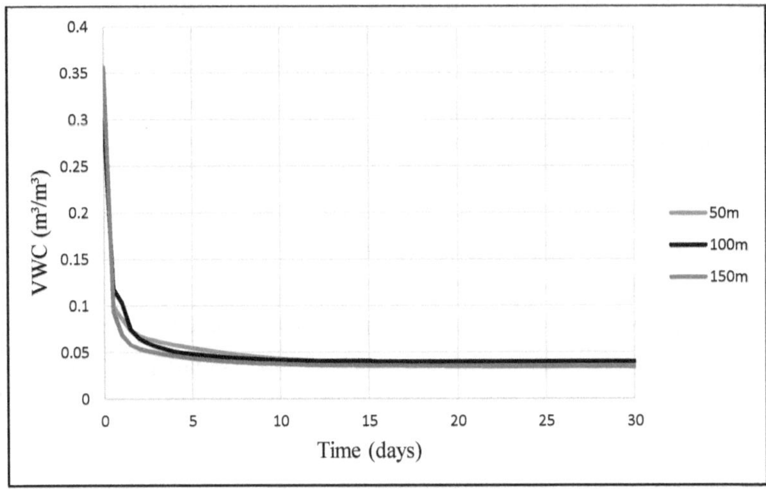

Figure 6.21 The variation of the volumetric water content at the middle of gravel layer

6.4.2 Pore water pressure changes over time

Figures 6.22 and 6.23 show the variations in the pore water pressure at the sand-gravel interface and in the middle of the coarse grained soil layer. These plots correspond to the same points used in the two figures (Figures 6.20 and 6.21) shown above.

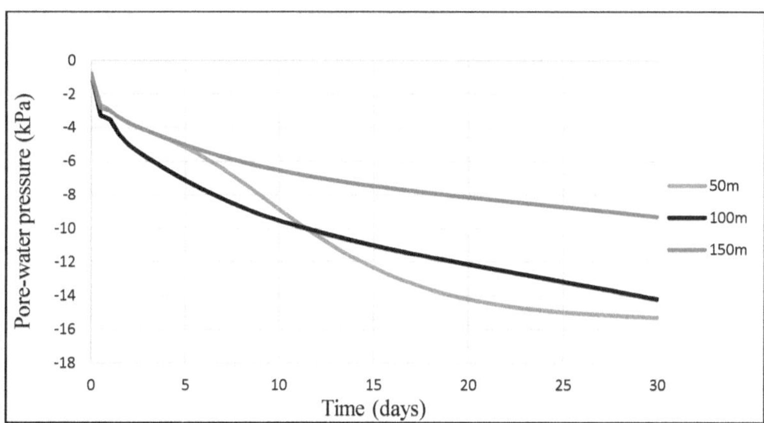

Figure 6.22 Variations in pore water pressure at sand-gravel interface

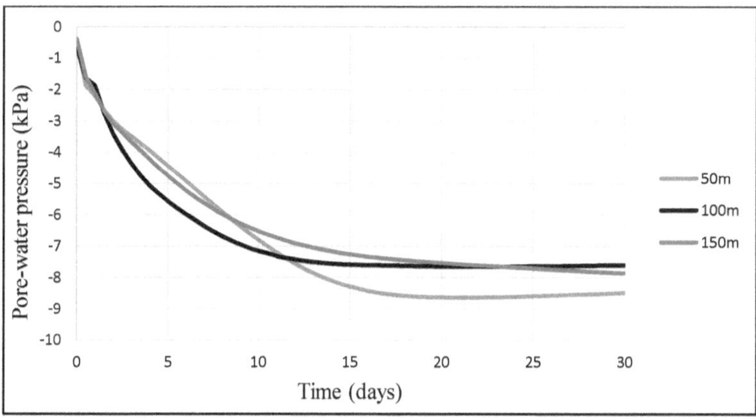

Figure 6.23 Variations in pore water pressure in the middle of gravel layer

A comparison between the heated air flow case (Figures 6.22 and 6.23) and without heated air flow case (Figures 6.11 and 6.12) shows that almost two to three fold increase in the matric suction takes place as a result of heated air flow.

FE analysis with heated airflow-360 days

A new analysis was carried out for a total duration of 360 days using the same mesh size and time steps. The air temperature boundary condition was added into the model. The temperature at all sides of the model was set at 24°C. The results of the FE analysis with heated air flow for a duration of 360 days are presented and discussed in terms of the changes in the contours of the VWC and matric suction.

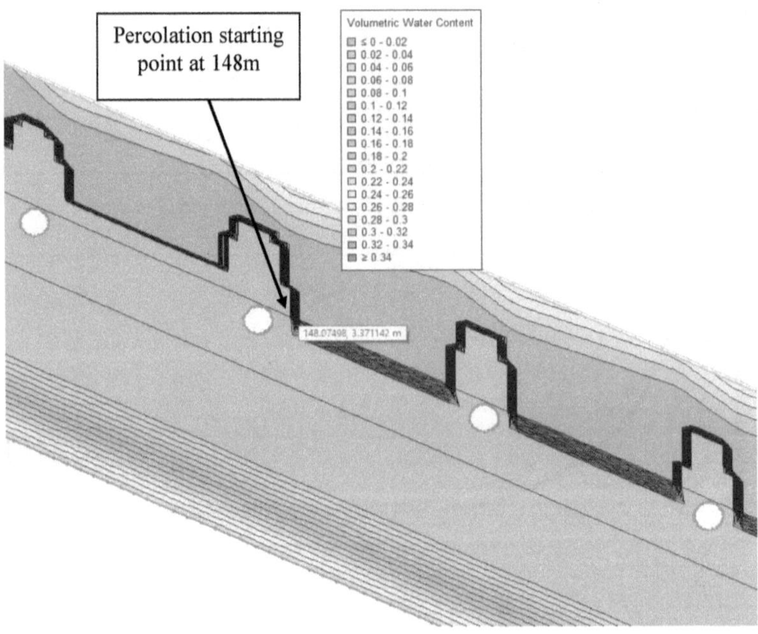

Figure 6.24 VWC changes in the soil layers resulting from heated air flow

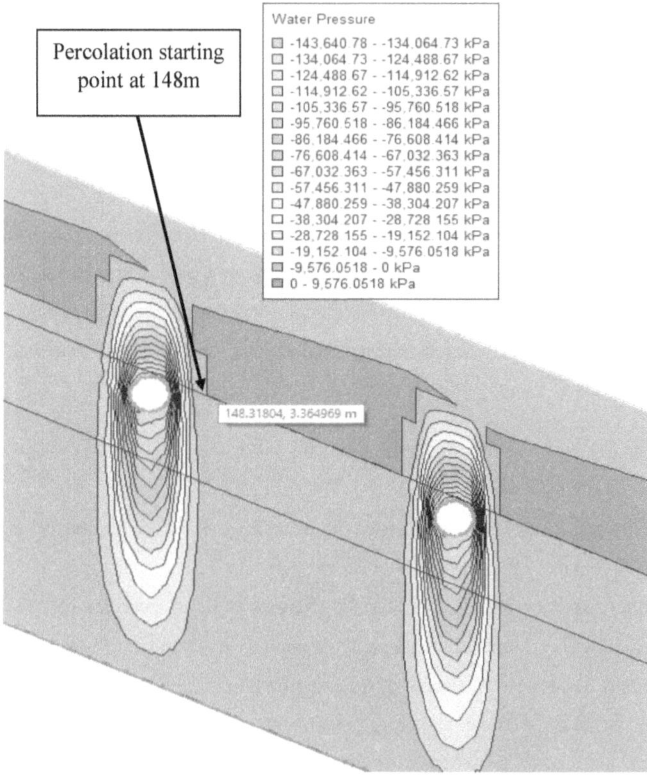

Figure 6.25 Water pressure changes in the soil layers resulting from heated air flow

6.5 Summary and discussion

The results presented in this chapter show that using the heated air flow has a positive influence on the behavior of the capillary barrier. The seepage into the municipal waste mass is better controlled. Soil moisture escapes faster through the pipes. The reduction in VWC means that the storage capacity of the system is improved in comparison with the landfill cover without heated air flow. It was noted that the diversion length of the capillary barrier system was increased using heated airflow from 19m (original research) to 148m for the same period of time.

CHAPTER 7

SUMMARY, CONCLUSIONS, AND RECOMMENDATIONS FOR FUTURE WORK

7.1 Summary

The specific objectives of this investigation were:

1- To develop a method that utilizes heated air flow to prevent infiltration of water into the coarse-grained soil layer of capillary barriers.

2- To determine the impact of heated air flow on the water removal rate and recovery of capillary barrier systems, which may have undergone a percolation breach event.

3- To conduct FEA to simulate the experimental study and to extend the applicability of this model to other case studies in order to demonstrate the benefits of heated air flow on the performance of capillary barrier systems.

To achieve the objectives of this study, the following tasks were undertaken:

1- A literature review was conducted.

2- A laboratory scale testing was carried out to determine the effect of heated air flow on the VWC and matric suction in a layer of soil representing the coarse grained soil layer of a potential capillary barrier (Chapter 3). Several types of instrumentation were used to measure the VWC, matric suction, and temperature at a number of locations.

3- A numerical analysis was conducted to simulate the behaviour of the soil mass subjected to thermal changes in the laboratory experiments.

4- Extend the method to two case studies to demonstrate the benefits of heated air flow in the coarse grained soil layer.

7.2 Conclusions

The method used in this investigation was based on: (1) application of temperature change at the perimeter of a pipe, installed in the coarse grained soil layer near the interface between the fine- and coarse-grained soil layers, and (2) application of suction as a boundary condition at the perforated parts of the pipe to decrease VWC and increase matric suction in the soil mass. Using this specific method, the results of the FE analyses of the laboratory experiments and the selected two case studies from the literature, showed that the heated air flow through the coarse grained soil layer of a capillary barrier would improve its performance as a soil cover in engineering practice. Comparisons of measured and calculated values of VWC and matric suction showed good agreement confirming the validity of the method. The model of the current study has been successfully applied to two case studies and the model predictions compare well with the experimental observations of the original researchers.

Based on the above conclusions, the author believes it would be possible to extend the benefits of employing heated airflow in capillary barriers as listed below.

1- For a given amount of infiltration, the time required for breakthrough can be extended a significant amount, as demonstrated in the case studies described in this investigation.

2- The amount of infiltration into the coarse-grained soil layer can be reduced, provided that all other boundary conditions and materials remain unchanged.

3- It becomes possible to use thinner soil layers to achieve the same capillary barrier effect.

4- For intensity of precipitation higher than the design precipitation, the capillary barrier can still continue to function as long as heated air flow is circulated in the system.

5- For a longer duration of the precipitation event, the same capillary barrier can continue to function as long as heated air flow is circulated in the system.

7.3 The major scientific contribution/ the contribution to knowledge

The current research provides a suitable, cost effective and long-term dewatering solution and extends the longevity of capillary barrier systems. Using preheated air flow in a capillary barrier removes the water from the system and as a result, water is prevented from reaching the fine-coarse interface (the downward movement of water will be limited) and the storage capacity of the system increases. Self-dewatering capillary barrier systems using preheated air flow can be implemented not only in civil engineering structures (e.g., road embankments) but also for environmental engineering applications (e.g., landfill covers).

The major contributions of the current research are:

- Building an instrumented sand box to study the effect of heated airflow on the performance of the coarse grained soil layer;

- Both the test results and the numerical analysis show that heated airflow substantially reduces the VWC and increases the matric suction in the soil. Therefore, heated airflow would improve the performance of capillary barriers significantly;
- A FEA was conducted using commercial software to simulate the behavior of the soil mass subjected to thermal changes. The applicability of the model was demonstrated in two case studies confirming the benefits of the heated airflow;
- A generated suction was applied as a boundary condition at the perforation opening in case study 1, and 2 and let the analysis provide the meaning of the results;
- The diversion length (DL) was increased/improved using heated airflow in the capillary barrier system, which means that the system can handle more water before breakthrough occurs (the system can function for longer periods).

The proposed research is expected to contribute to both geotechnical and environmental engineering practice. The developed framework can find various applications, such as slope stabilization techniques to prevent slope failure for road embankments and landfill cover systems using natural processes (i.e., using sun thermal energy to generate airflow). While practical means of generating the heated airflow in the field remains an open question, passive solar systems could be potentially used to achieve this goal.

7.4 Limitations

The limitations of the current study are;

1. During the experimental process, no suitable copper pipe diameter is found for the condensation test.

The condensation test was performed to obtain a measure of the moisture removed from the sand as water vapor in the air flowing through the buried pipe. The test was prepared using a coil copper pipe (air condenser pipe) submerged in a cooling tank containing cold water. The author of the current book proposes the following mechanism of moisture removal from the sand. The hot air flowing inside the buried pipe has the ability to attract moisture (through vacuum process) from the sand in the box, and convert it into water vapour due to the effect of airflow and heat inside the pipe. The condensation test was run for two days but it was discontinued after that initial period due to unexpected performance. It was observed that instead of freely flowing through the condenser pipe, the air was forced to circulate within the sand box. This was due to the selected copper pipe diameter (1 inch) being too small to pass the airflow exiting the pipe (2 inches) and was instead obstructing the flow. The friction in the copper pipe (30 feet in length) caused an additional decrease in the airflow rate. Figure 7.1 shows the test set up for the condensation test. Figure 7.2 shows the condenser pipe placed in the cooling water tank. Figure 7.3 shows the effect of the return airflow in the sand box caused by the factors mentioned above.

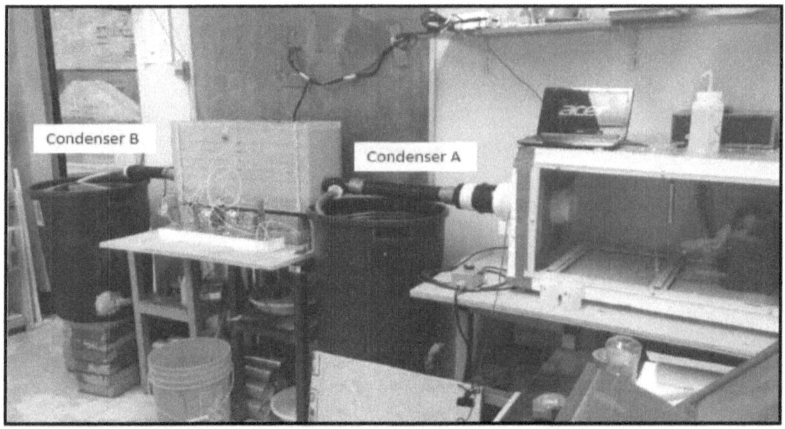

Figure 7.1 The condensation test set up

Figure 7.2 The Condenser -coil pipe

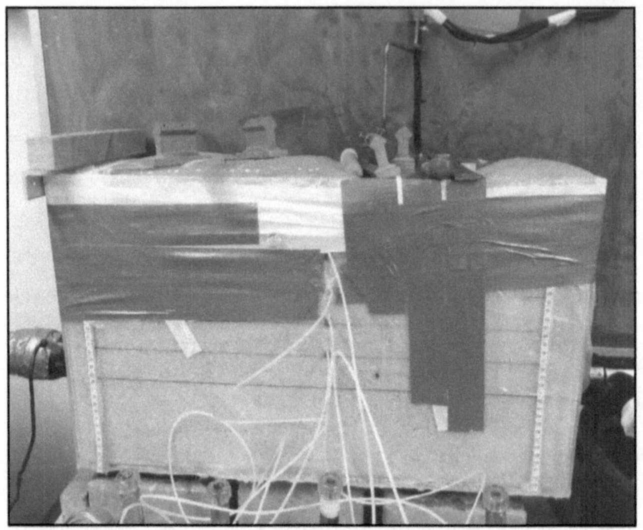

Figure 7.3 The puffiness of the plastic sheet cover

2. The laboratory experimental tests were not extended to field testing to study the effect of the heated airflow on the drying process in the field.

3. The effect of heated airflow on the mechanical behavior of the soil mass was not considered and, in particular, no investigation of the effect of the approach on the shear strength and slope stability was done. It was not possible to perform an analysis using Slope/W (for slope stability analysis), and Temp/W in combination to study the effect of heated airflow on the slope stability (the capability of the software to perform such analysis is limited).

7.5 Recommendations for future work

The following studies will be beneficial improvements of the performance of capillary barriers where the heated air flow method is employed.

- Study the effect of the number of pipes used in the system;

- Study the effect of the diameter of pipes;

- Study the effect of the perforation size in the pipes;

- Study the effect of the number perforations in the pipe wall;

- Study the effect of different heated air flow rate;

- Study the effect of different temperature on the water removal rate;

- Study the effect of the compaction on the performance of the system;

- Study the effect of heated airflow on the drying process in the field;

- Study the effect of heated airflow on the mechanical behavior of the system.

In the present study, no attempt was made to calculate the cost of including a heated air flow system in the construction of a capillary barrier. A study to that effect would be highly desirable.

REFERENCES

Abdolahzadeh, A.M., Lacroix Vachon, B., and Cabral, A.R. 2008. *Hydraulic barrier and its impact on the performance of cover with double capillary barrier effect.* In Proceedings of the 61st Canadian Geotechnical Conference, Edmonton, 21–24 September.

Abdolahzadeh, A.M., Vachon, B.L., and Cabral, A.R. 2011a. *Evaluation of the effectiveness of a cover with capillary barrier effect to control percolation into a waste disposal facility.* Canadian Geotechnical Journal, 48(7): 996– 1009. doi:10.1139/t11-017.

Abdolahzadeh, A.M., Vachon, B.L., and Cabral, A.R. 2011b. *Assessment of the design of an experimental cover with capillary barrier effect using 4years of field data.* Geotechnical and Geological Engineering, 29(5): 783–802. doi:10. 1007/s10706-011-9417-x.

Abu-Hamdeh.N.H. 2003. *Thermal Properties of Soils as affected by Density and Water Content.* Biosystems Engineering 86 (1), 97–102 doi: 10.1016/S1537-5110(03)00112-0. SW-Soil and Water.

Abu-Zreig.M.M. Abe. Y. and Isoda. H. 2006. *Study of salt removal with evaporation drainage method. Canadian biosystems engineering.* Volume 48:1.25 - 1.30.

Abu-Zreig. M. Abe.Y. Isoda.H., and Abo-Izreik. A. 2007. *Land Drainage with Evaporation Drainage Method.* ASABE Annual International Meeting. Paper Number: 072189. Minneapolis, Minnesota. USA

Abo-Izreik. A., and Abu-Zreig. M .2012. *Soil Reclamation and Drainage with Evaporation Drainage Methods.* ASABE Annual International Meeting. Paper Number: 12-1340963. Dallas, Texas. USA

Adala.M. Bennacer.R. Sammouda. H., and Guizani. A. 2009. *Heat and Mass Transport in the Unsaturated Porous Media: Application to the Soil Dry Drainage Method.* Defect and Diffusion Forum Vols. 283-286 (2009) pp 589-593. Trans Tech Publications, Switzerland doi:10.4028/www.scientific.net/DDF.283-286.58.

Ahmed E. Ahmed, M. Y. Hamdy, W. L. Roller, D. L. Elwell. *Technical feasibility of utilizing reject heat from power stations in greenhouses*. ASAE, 1982.

Akhter.M.G. Ahmad. Z., and Khan.K.A. 2006. *Excel based finite difference modeling of ground water flow*. Journal of Himalayan Earth Sciences 39, 49-53

Aubertin, M. Cifuentes, E. Apithy, S. A. Bussière, B. Molson, J., and Chapuis, R.P. 2009. *Analyses of water diversion along inclined covers with capillary barrier effects*. Canadian Geotechnical Journal. 46(10): 1146-1164, 10.1139/T09-050.

ASTM (1998). American Society for Testing and Materials, *Standard practice for classification of soils for engineering purposes* (Unified Soil Classification System), ASTM D2487-98.

ASTM (1998). American Society for Testing and Materials, *Standard test methods for specific gravity of soil solids by water pycnometer*, ASTM D854-98.

ASTM (2004). American Society for Testing and Materials, *Standard test for sieve analysis*, ASTM D6913-04

ASTM (2006a). American Society for Testing and Materials, *Standard test method for permeability of granular soils (constant head)*, ASTM D2434–68.

ASTM (2008). American Society for Testing and Materials, Standard test methods for Determination of the Soil Water Characteristic Curve for Desorption Using a Hanging Column, Pressure Extractor, Chilled Mirror Hygrometer, and/or Centrifuge, ASTM D6836-02

ASTM (2010). American Society for Testing and Materials, *Standard test methods for laboratory determination of water (moisture) content of soil and rock by mass*, ASTM D2216-10

ASTM (2016). American Society for Testing and Materials, *Standard test method for Minimum index density and unit weight of soils and calculation of relative density*, ASTM D4254-16.

ASTM (2016). American Society for Testing and Materials, *Standard test methods for maximum index density and unit Weight of soils using a vibratory table*, ASTM D4253–16

Bach. L. B.1992. *Soil Water Movement in Response to Temperature Gradients: Experimental Measurements and Model Evaluation.* The Soil Science Society of America Journal, Volume 56. No. 1.

Baehr. A. L. Hoag. G. E., and Marley. M. C. 1989. *Removing volatile contaminants from the unsaturated zone by inducing advective air-phase transport.* Journal of Contaminant Hydrology, 4 .1261 Elsevier Science Publishers B.V., Amsterdam.

Bahador.M. Evans. T.M. and Gabr. M.A. 2014. *Numerical studies on the effect of temperature on the unsaturated hydraulic response of geotextiles.* Computers and Geotechnics 59 (2014) 161–170.

Benson, C. H., Bosscher, R. J. 1999. *Time-Domain Reflectometry (TDR) in Geotechnics: A Review.* Non-destructive and Automated Testing for Soil and Rock Properties, ASTM STP 1350, Marr W. A., and Fairhurst. C. E. Eds., American Society for Testing and Materials, West Conshohocken, PA.

Besser. R.S. 2002. *Spreadsheet solutions to two-dimensional heat transfer problems.* ChE division of ASEE. Pp 160-165.

Bittelli.M. 2010. *Measuring Soil Water Potential for Water Management inAgriculture: A Review.* Sustainability ISSN 2071-1050

Brian. A. A., and Craig. H. B. 2002. *Predicting Airflow Rates in the Coarse Layer of Passive Dry Barriers. Journal of geotechnical and geoenvironmental engineering.* ASCE, ISSN 1090-0241/2002/4-338–346.

British Columbia Ministry of Environment .2016. *Landfill Criteria for Municipal Solid Waste. Second Edition.*

Boulard, T. Razafinjohany, E., and Baille, A. 1989. *Heat and vapour transfer in a greenhouse with an underground heat storage system (Parts 1) experimental test* .Agriculture Forest Meteorology. Vol 45, Issues 3-4, pgs. 175-184.

Boulard, T. Razafinjohany, E., and Baille, A. 1989. *Heat and vapour transfer in a greenhouse with an underground heat storage system (Parts 2) Model.* Agriculture Forest Meteorology.Vol 45, Issues 3-4, pgs. 185-194.

Bussiere, B. Apithy, S. Aubertin, M., and Chapuis, R. 2003. *Water diversion capacity of inclined capillary barriers*. In Proceedings of the 56th Canadian Geotechnical Conference, Winnipeg. Canada.

Bussiere, B. Aubertin, M., and Chapuis, R. 2000. *An Investigation of slope effect on the efficiency of capillary barrier to control AMD*. In Proceedings of the Fifth International Conference on Acid Rock Drainage, Denver, CO, May 21-24, pp. 969-977.

Carpena.R. M. Shukla.S.,and Morgan.K. 2004. *Field Devices for Monitoring Soil Water Content.* U.S. Department of Agriculture under Agreement No. 2004-51130-03114.

Dakshanamurthy. V., and Fredlund. D. G.1980. *Moisture and air flow in an unsaturated soil.* Proceedings of the fourth international conference on expansive soils, vol.1, pp. 514-532. Denver, Colorado.

Dakshanamurthy. V., and Fredlund. D. G.1981. *A mathematical model for predicting moisture flow in an unsaturated soil under hydrolic and temperature gradients.* The emergence of unsaturated soil mechanics 687-697.

Dakshanamurthy. V., and Fredlund. D. G.1982. *Transient flow process in unsaturated soils under flux boundary condition.* Proceeding of the fourth international conference on numerical method in geomechanics, Edmonton.

Dalton, F.N., Van Genuchten, M. Th. 1986. *The time-domain reflectometry method for measuring soil water content and salinity.* Elsevier Science Publishers B.V., Amsterdam. Geoderma, 38 (1986) 237-250.

Decagon Devices Inc. 2016a. *MPS-2 & MPS-6 Dielectric Water Potential Sensors. Operator's Manual.* Web 15 May. 2017. www.decagon.com.

Decagon Devices Inc. 2016b. *Water Content, Electrical Conductivity, and Temperature sensor (5TM) user Manual.* Web 15 May. 2017. www.decagon.com

Edwards. R., and Lobaugh. M. 2014. *Using Excel to Implement the Finite Difference Method for 2-D Heat Transfer in a Mechanical Engineering Technology Course.* American Society for Engineering Education.

El-Keshky. M. 2011. *Temperature Effect on the Soil Water Retention Characteristic*. Master book. Arizona state university.USA.

Elwell, D. L. Roller, W. L., and Ahmed. A. E. 1982. *Waste Heat Utilization for Soil Heating in Greenhouses*. American Society of Agricultural Engineers.

Esmail M. A. Mokheimer, and Mohamed A. Antar. 1998. *on the use of spreadsheets in heat conduction analysis*. International Journal of Mechanical Engineering Education Vol 28 No 2.

Fredlund, D. G. Rahardjo, H., and Fredlund, M. D. 2012. *Unsaturated Soil Mechanics in Engineering Practice* (chap 4.P109-183) John Wiley & sons, Inc. ISBN: 9781118280492.

Fredlund, D. G. Rahardjo, H., and Fredlund, M. D. 2012a. *Unsaturated Soil Mechanics in Engineering Practice* (chap 5.P184-272) John Wiley & sons, Inc. ISBN: 9781118280492.

Fredlund, D. G. Rahardjo, H., and Fredlund, M. D. 2012b. *Unsaturated Soil Mechanics in Engineering Practice* (chap 9.P450-486) John Wiley & sons, Inc. ISBN: 9781118280492.

Fredlund, D. G. Rahardjo, H.,and Fredlund, M. D. 2012c. *Unsaturated Soil Mechanics in Engineering Practice* (chap 10.P487-519) John Wiley & sons, Inc. ISBN: 9781118280492.

Fredlund, D. G., Wilson, G. W. Clifton. A. W., and Barbour. S. L. 1999. *The emergence of unsaturated soil mechanics: Fredlund volume*. Ottawa: NRC Research Press.

Fredlund, D.G. and Xing, A. (1994). *Equations for the soil-water characteristic curve*, Canadian Geotechnical Journal, 31(4): 521-532.

Fredlund, D.G., Xing, A. and Huang, S. (1994). *Predicting the permeability function for unsaturated soils using the soil-water characteristic curve*, Canadian Geotechnical Journal, 31(4): 533-546.

Fredlund, M. D. Zhang, J. M. Tran, D., and Fredlund. D.G. 2011. *Coupling Heat and Moisture Flow for the Computation of Actual Evaporation*. Pan-Am CGS Geotechnical Conference.

Gurr. C.G. Marshall. T. J. Hutton .J. T. 1952. *Movement of water in soil due to a temperature gradient.* Soil science volume 74. Number 5. Commonwealth scientific and industrial research organization, Adelaide. Australia

Geo-Slope International Ltd. 2014. Vadose Zone Modeling with VADOSE/W. An Engineering Methodology. Web: http://www.geo-slope.com

Haigh. S.K. 2012. *Thermal conductivity of sands.* Ge´otechnique [http://dx.doi.org/10.1680/geot.11.P.043]

Hamdhan. I. N., and Clarke. B. G. *Determination of Thermal Conductivity of Coarse and Fine Sand Soils.* Proceedings World Geothermal Congress 2010 Bali, Indonesia, 25-29 April 2010

Hollmuller, P. Lachal, B. 2000. *Cooling and preheating with buried pipe systems: monitoring, simulation and economic aspects.* Energy and Buildings 1295.

Hollmuller, P. Lachal, B. 2005. *Buried pipe systems with sensible and latent heat exchange: validation of numerical simulation against analytical solution and long-term monitoring.* 9th Conference of international building performance simulation association.

Holman. J.P.1986. Heat Transfer. Six edition. ISBN 0-07-Y66459-5.

Interstate Technology & Regulatory Council (ITRC). 2003. *Technology Overview Using Case Studies of Alternative Landfill Technologies and Associated Regulatory Topics.* ALT-1. Web 5 April. .2020. http://www.itrcweb.org/ALT-1.pdf.

Karimi.A. 2008. *Use of Spreadsheets in Solving Heat Conduction Problems in Fins.* American Society for Engineering Education.

Karimi.A. 2009. *A compilation of examples for using excel in solving heat transfer problems.* University of Texas. American Society for Engineering Education. American Society for Engineering Education. Pp 14.17.1 -14.17.42.

Kaye. G. W. C., and Laby. T. H. 1911. *Tables of physical and chemical constants and some mathematical functions.* Longmans, green, and co. page 40.

Khire, M. V. Benson, C. H., and Bosscher, P. J. 2000. *Capillary Barriers: Design Variables and Water Balance*. Journal of Geotechnical and Geoenvironmental Engineering, ASCE, Vol. 126, No. 8, pgs. 695-708.

Krisdani, H. 2006. *The Use of Residual Soil and Geosynthetic Material in Capillary Barrier System*. PhD book, School of Civil and Environmental Engineering. Nanyang Technological University, Singapore.

Krisdani, H. Rahardjo, H., and Leong, E.C. 2005. *Behaviour of Capillary Barrier System Constructed using Residual Soil*. Geo-Frontiers Congress. ASCE. Austin, Texas, USA, Issue 130-142, pp. 3525-3539

Li.Y. and Luo Z. X.2000. *Physical Mechanisms of Moisture Diffusion into Hygroscopic Fabrics during Humidity Transients*. Journal of the Textile Institute, 91:2, 302-316, DOI: 10.1080/00405000008659508

Lier. Q.J.V., and Durigon. A. 2012. *Soil thermal diffusivity estimated from data of soil temperature and single soil component properties*. R. Bras. Ci. Solo, 37:106-112

Lindblom. J. 2007. *Condensation Irrigation simulations of heat and mass transfer*. Licentiate book 2006:08, Luleå University of Technology, Sweden, 2006.434J. Lindblom, B. Nordell / Desalination 203 (2007) 417–434

Lindblom. J., and Nordell, B. 2006. *Condensation Irrigation system for desalination and irrigation*. Submitted to Transport in Porous Media.

Lindblom. J., and Nordell, B. 2006. *Water Productionby Underground Condensation of Humid Air*. Desalination 189 (2006) 248–260.

Lindblom. J., and Nordell, B.2012. *Experimental study of underground irrigation by condensation of humid air in perforated pipes*. Technical report. ISSN: 1402-1536. ISBN 978-91-7439-513-6

Lu. N., and Likos. W. 2004a. *Unsaturated soil mechanics*. (Chap 8.P325-368) John Wiley & sons, Inc. ISBN: 0-471-44731-5

Lu. N., and Likos. W. 2004b. *Unsaturated soil mechanics*. (Chap 9.P369-413) John Wiley & sons, Inc. ISBN: 0-471-44731-5

Mahmud. M. 1996. *Spreadsheet solutions to Laplace's equation: seepage and flow net*. Journal Technology. University Technology Malaysia. Vol. 25. Pp 53- 67.

Mancarella.D, and Simeone.V. 2012. *Capillary barrier effects in unsaturated layered soils, with special reference to the pyroclastic veneer of the Pizzo d'Alvano*, Campania, Italy Bull Eng Geol Environ (2012) 71:791–801 DOI 10.1007/s10064-012-0419-6

Menziani, M., Rivasi, M.R., Pugnaghi, S., Santangelo, R., and Vincenzi, S. 1996. *Soil volumetric content measurements using TDR technique*. Annali Di Geofisica VOL. XXXIX, N. Italy.

Morris, C.E. 1999. *Moisture Removal from a Two-Layer Porous Media: A Conceptual Model and Experimental Result*. Kluwer Academic Publishers. Transport in Porous Media 36: 23– 42.

Morris, C.E., and Stormont. J. C. 1999. *Parametric study of unsaturated drainage layers in a capillary barrier*. Journal of Geotechnical and Geoenvironmental Engineering

Morris, C.E. Thomson, B. M., and Stormont, J. C. 1999. *Design of dry barriers for containment of contaminants in unsaturated soils*. Groundwater Monitoring and Remediation journal. Volume 19, Issue 3.

OMEGA engineering Inc .*Humidity and Temperature Controllers*. Web. March 9. 2016a. http://www.omega.com/temperature/pdf/ITH.pdf .

OMEGA engineering Inc. *Mini Van Anemometer (HHF801)*. User Manual. Web. March 9. 2016b. https://www.omega.com/manuals/manualpdf/M4793.pdf

Pachepsky. Y. Timlin. D., and Rawls. W. 2003. *Generalized Richards' equation to simulate water transport in unsaturated soils*. Journal of Hydrology 272, 3–13.USA

Philip. J.R., and De Vries. D. A. 1957. *Moisture Movement in Porous Materials under Temperature Gradients*. Transactions, American Geophysical Union. Volume. 38, No. 2.

Pinel. P., and Morrison.I.B. 2012. *Coupling soil heat and mass transfer models to foundations in whole-building simulation packages*. Proceedings of eSim 2012: The Canadian Conference on Building Simulation.

Rahardjo, H. Krisdani, H., and Leong, E.C. 2007. *Application of unsaturated soil mechanics in capillary barrier system*. Proceedings of 3rd Asian Conference on Unsaturated Soils. Nanjing, China, 21–23 April, Science Press. pp. 127 –137.

Rahardjo, H. Krisdani, H. Leong, E.C. Ng, Y.S. Foo, M.D., and Wang, C.L. 2007e. *Capillary Barrier as Slope Cover*. Proceedings of 10th Australia New Zealand Conference on Geomechanics "Common Ground". Brisbane, Australia, 21–24 October, Vol.2, pp. 698–703.

Rahardjo, H. Rezaur, R.B. and Leong, E.C. 2009. *Mechanism of rainfall-induced slope failures in tropical regions*. In Proceedings of the 1st Italian Workshop on Landslides (IWL) Rainfall-induced landslides; Mechanisms, Monitoring Techniques and Nowcasting Models for Early Warning System. Department of Civil Engineering, Seconda University. Naples. Italy. 1:31-42.

Rahardjo. H. Satyanaga.A., and Leong. E. C. 2012. *Unsaturated Soil Mechanics for Slope Stabilization*. Geotechnical Engineering Journal of the SEAGS & AGSSEA Vol. 43 No.1 ISSN 0046-5828

Recktenwald. G. W. 2011. *Finite Difference Approximations to the Heat Equation*. Portland, Oregon, USA

Rollins. R. L. 1954. *Movement of soil moisture under a thermal gradient*. Retrospective Theses and Dissertations. 13247. http://lib.dr.iastate.edu/rtd/13247

Roshani. P.2014. *The Effect of Temperature on the SWCC and Estimation of the SWCC from Moisture Profile under a Controlled Thermal Gradient*. Master book. University of Ottawa.

Sakai.M. Toride.N.,and Šimůnek.J. 2009. *Water and Vapor Movement with Condensation and Evaporation in a Sandy Column*. SSSAJ: Volume 73: Number 3. Soil Sci. Soc. Am. J. 73:707-717. doi:10.2136/sssaj2008.0094

Salah. M. Infante.J.A., and Dimitrova.R. *Self-dewatering capillary barrier system (SDCBS) using preheated air flow method* .In proceeding, 70th Canadian Geotechnical

Conference and the 12th Joint CGS/IAH-CNC Groundwater Conference, 1-4 October 2017, Ottawa, ON.

Schumack.M. 2004. *Use of a Spreadsheet Package to Demonstrate Fundamentals of Computational Fluid Dynamics and Heat Transfer*. Int. J. Engng Ed. Vol. 20, No. 6, pp. 974±983. Tempus Publications.

Smits. K.M. Cihan. A. Sakaki.T, and Illangasekare. T. H. 2011. *Evaporation from soils under thermal boundary conditions: Experimental and modeling investigation to compare equilibrium and non-equilibrium-based approaches.* Water resources research, vol. 47, w05540, doi: 10.1029/2010wr009533.

Song, W.K. Cui Y.J. Tang, A.M., and Ding. W.Q. 2015. *Water evaporation experiments in environmental chamber.* Pp 1359-1364. Computer Methods and Recent Advances in Geomechanics. Taylor & Francis Group, London, ISBN 978-1-138-00148-0

Stormont, J.C. 1996. *The Effectiveness of Two Capillary Barriers on a 10% Slope*. Geotechnical and Geological Engineering, 14, pp. 243-267.

Stormont, J. C. Ankeny, M. D. Burkhard, M.E. Tansey, M.K., and Kelsey. J.A. 1994. *Assessment of an active dry barrier for a landfill cover system.* US department of energy office of technology development. SAND94-0301.

Stormont, J. C. Ankeny, M. D., and Kelser, J. A. 1998. *Airflow as monitoring technique for landfill liners.* Journal of environmental engineering. vol. 124, No.6. ASCE, ISSN 0733-9372/98/00060539.

Stormont, J. C. Ankeny, M. D. Tansey, M.K., and Daniel B. Stephens. 1994. *Water removal from a dry barrier cover system*. SAND94-0676.

Tami, D. 2003. *Mechanism of Sloping Capillary Barriers under High Rainfall Conditions*. PhD book, School of Civil and Environmental Engineering, Nanyang Technological University, Singapore.

Tami, D. Rahardjo, H. Leong, E.C., and Fredlund, D.G. 2004a. *Design and laboratory verification of a physical model of sloping capillary barrier*. Canadian Geotechnical Journal. 41: 814–830.

Tami, D. Rahardjo, H. Leong, E.C., and Fredlund, D.G. 2004b. *A physical model for sloping capillary barrier*. Geotechnical Testing Journal. 27(2): 1–11.

Taylor. S., and Cavazza.L. 1954. *The Movement of Soil Moisture in Response to Temperature Gradients*. Division I-Soil physics. Soil science society of America proceedings. VOL. 18. No. 4.

TE Connectivity Company. 2015. HTM2500LF – Temperature and Relative Humidity Module. Data sheet. Web. Nov 9. 2016. http://www.te.com

Tidwell, V. C., Glass, R.J., Chocas, C., Barker, G., Orear, L. 2003. *Visualization experiment to investigate capillary barrier performance in the context of a Yucca Mountain emplacement drift*. Journal of contaminant hydrology. 62-63:287-301.

Tokoro. T. Ishikawa.T. Shirai.S. and Nakamura.T. 2016. *Estimation methods for thermal conductivity of sandy soil with electrical characteristics*. The Japanese Geotechnical Society, Soils and Foundations 2016; 56(5):927–936. http://dx.doi.org/10.1016/j.sandf.2016.08.016.

Topp, G. C., Davis J. L., Annan .A. P. 1980. *Electromagnetic determination of soil water content: measurements in coaxial transmission lines*. Water resources research, vol. 16, no. 3, pages 574-582.

Tripathy.S. Al-Khyat. S. Cleall. P.J. Baille.W., and Schanz. T. 2016. *Soil Suction Measurement of Unsaturated Soils with a Sensor Using Fixed-Matrix Porous Ceramic Discs*. Indian Geotech Journal. 46(3):252–260

The Ministry of Environment and Energy.2012. *A guideline on the regulatory and approval requirements for new or expanding landfilling sites*. Landfill Standards. A publication of Government of Ontario. Document: 331938.

The Minister of Justice. 1999. *Canadian Environmental Protection Act.* Web 5 April. .2020. http://laws-lois.justice.gc.ca

U.S. Environmental Protection Agency (EPA). 1989a. *Final Covers on Hazardous Waste Landfills and Surface Impoundments*. Technical Guidance Document, EPA/530-SW-89-047. Office of Solid Waste and Emergency Response. Washington, D.C.

U.S. Environmental Protection Agency (EPA). 2004. *Technical Guidance for RCRA/CERCLA, Final Covers.* Document, EPA/540-R-04-007. Office of Solid Waste and Emergency Response. Washington, D.C.

Vachon. B. L., Abdolahzadeh, A. M., and Cabral. A. R. 2015. *Predicting the diversion length of capillary barriers using steady state and transient state numerical modeling: case study of the Saint-Tite-des-Caps landfill final cover Benoit Lacroix.* Canadian Geotechnical Journal, 52: 2141–2148 (2015) dx.doi.org/10.1139/cgj-2014-0353

Vanapalli. S.K. Nicotera. M. V. and Sharma. R.S. 2008. Axis Translation and Negative Water Column Techniques for Suction Control. Journal Geotechnical and Geological Engineering, Volume 26, No. 6, 645–660. DOI: 10.1007/s10706-008-9206-3

Vanapalli. S.K. Salinas. L.M. Avila. D., and Karube. D. 2002. *Suction and storage characteristics of unsaturated soils.* The 3rd International Conference of Unsaturated Soils. Recife, Brazil.

Van Genuchten. M. T. 1980. A Closed-form Equation for Predicting the Hydraulic Conductivity of Unsaturated Soils. Soil Science Society of America Journal 44(5) DOI: 10.2136/sssaj1980.03615995004400050002x

Vivancos. V. L., and Voller. V.R. 2004. *Two Numerical Methods for Modeling Variably Saturated Flow in Layered Media.* Published in Vadose Zone Journal 3:1031–1037.USA

Wallen. B.M. Smits.K.M. Sakaki.T. Howington. S.E., and Chamindu Deepagoda T.K.K. *Thermal Conductivity of Binary Sand Mixtures Evaluated through Full Water Content Range.* Soil Physics & Hydrology. Soil Sci. Soc. Am. J. 80:592–603 doi:10.2136/sssaj2015.11.0408.

Wilson, G.W. Barbour, S.L., and Fredlund, D.C. 1995. *The prediction of evaporative fluxes from unsaturated soil surfaces.* ISBN 9054105836/ 2859782419.

Wilson. G. W. Fredlund. D.G., and Barbour. S.L.1994. *Coupled soil-atmosphere modelling for soil evaporation.* Can. Geotech. J. 31, 151-161.

Wilson, G.W. Fredlund, D.G., and Barbour, S.L. 1997. *The effect of soil suction on evaporative fluxes from soil surfaces.* Canadian Geotechnical Journal. 34: 145–155.

Yang, H. Rahardjo, H. Leong, E.C., and Fredlund, D.G. 2004. *A study of infiltration on three sand capillary barriers*. Canadian Geotechnical Journal, 41: 629–643.

Yang.H. Rahardjo.H. Leong.E. C., and Fredlund. D.G. 2004. *Factors affecting drying and wetting soil-water characteristic curves of sandy soils*. Can. Geotech. J. 41: 908–920. doi: 10.1139/T04-042

Yun T. S., and Santamarina J. C. 2008. *Fundamental study of thermal conduction in dry soils*. Granular matter 10:197–207 DOI 10.1007/s10035-007-0051-5

Zhai. Q. Rahardjo. H. 2012. *Determination of soil–water characteristic curve variables*. Computers and Geotechnics 42 (2012) 37–43. doi:10.1016/j.compgeo.2011.11.010

Zhang.N. and Wang. W. 2017. *Review of soil thermal conductivity and predictive models*. International Journal of Thermal Sciences 117 (2017) 172e183. http://dx.doi.org/10.1016/j.ijthermalsci.2017.03.013

9 783384 220943